Advanced Maths Essentials
Core 3 for Edexcel

Welcome to Advanced Maths Essentials: Core 3 ... improve your examination performance by fo... ...ils you will need in your Edexcel Core 3 examination. It has been divi... ...y chapter into the main topics that need to be studied. Each chapter has then been divided by sub-headings, and the description below each sub-heading gives the Edexcel specification for that aspect of the topic.

The book contains scores of worked examples, each with clearly set-out steps to help solve the problem. You can then apply the steps to solve the Skills Check questions in the book and past exam questions at the end of each chapter. If you feel you need extra practice on any topic, you can try the Skills Check Extra exercises on the accompanying CD-ROM. At the back of this book there is a sample exam-style paper to help you test yourself before the big day.

Some of the questions in the book have a  symbol next to them. These questions have a PowerPoint® solution (on the CD-ROM) that guides you through suggested steps in solving the problem and setting out your answer clearly.

Using the CD-ROM

To use the accompanying CD-ROM simply put the disc in your CD-ROM drive, and the menu should appear automatically. If it doesn't automatically run on your PC:

1. Select the My Computer icon on your desktop.
2. Select the CD-ROM drive icon.
3. Select Open.
4. Select core3_for _edexcel.exe.

If you don't have PowerPoint® on your computer you can download PowerPoint 2003 Viewer®. This will allow you to view and print the presentations. Download the viewer from http://www.microsoft.com

Pearson Education Limited
Edinburgh Gate
Harlow
Essex
CM20 2JE
England
www.longman.co.uk

First published 2005
10 9 8 7 6 5
ISBN   978-0-582-83669-3

Design by Ken Vail Graphic Design

Cover design by Raven Design

Typeset by Tech-Set, Gateshead

Printed in Great Britain by Ashford Colour Press Ltd

The publisher's policy is to use paper manufactured from sustainable forests.

The publisher wishes to draw attention to the Single-User Licence Agreement at the back of the book. Please read this agreement carefully before installing and using the CD-ROM.

The Publisher and Authors would like to thank Rosemary Smith for her significant contributions to Chapters 1 and 5 of this book.

We are grateful for permission from the London Qualifications Limited trading as Edexcel to reproduce past exam questions. All such questions have a reference in the margin. London Qualifications Limited trading as Edexcel can accept no responsibility whatsoever for accuracy of any solutions or answers to these questions.

Every effort has been made to ensure that the structure and level of sample question papers matches the current specification requirements and that solutions are accurate. However, the publisher can accept no responsibility whatsoever for accuracy of any solutions or answers to these questions. Any such solutions or answers may not necessarily constitute all possible solutions.

1 Algebra and functions

1.1 Simplifying rational expressions

Simplification of rational expressions including factorising and cancelling, and algebraic division.

To **simplify** an algebraic fraction:

- Fully factorise the numerator and the denominator.
- Cancel common factors.

Example 1.1 Simplify $\dfrac{x^2 - 2x}{3x^2 - 7x + 2}$.

Step 1: Factorise both the numerator and the denominator fully.

Step 2: Cancel common factors.

$$\frac{x^2 - 2x}{3x^2 - 7x + 2} = \frac{x(x - 2)}{(3x - 1)(x - 2)}$$

$$= \frac{x(x - 2)^1}{(3x - 1)(x - 2)^1}$$

$$= \frac{x}{3x - 1}$$

Tip:
Do not attempt any cancelling until the expressions are in factorised form.

Note:
This *cannot* be simplified further.

To **multiply** algebraic fractions:

- Fully factorise all numerators and denominators.
- Cancel common factors.
- Multiply the numerators and multiply the denominators.

To **divide** algebraic fractions:

- Change $\div$ to $\times$ and invert the fraction immediately after the $\div$ sign.
- Follow the steps for multiplication.

Example 1.2 Express $\dfrac{4y^2 - 9}{2y^2 + 13y + 15} \div \dfrac{2y^2 - 3y}{y^2}$ as a single fraction in its simplest form.

Step 1: Change $\div$ to $\times$ and invert the second fraction.

Step 2: Factorise the numerators and denominators fully.

Step 3: Cancel common factors.

Step 4: Multiply the numerators and multiply the denominators.

$$\frac{4y^2 - 9}{2y^2 + 13y + 15} \div \frac{2y^2 - 3y}{y^2} = \frac{4y^2 - 9}{2y^2 + 13y + 15} \times \frac{y^2}{2y^2 - 3y}$$

$$= \frac{(2y + 3)(2y - 3)}{(2y + 3)(y + 5)} \times \frac{y^2}{y(2y - 3)}$$

$$= \frac{{}^1(2y + 3)(2y - 3)^1}{{}_1(2y + 3)(y + 5)} \times \frac{{}^1y \times y}{y(2y - 3)^1{}_1}$$

$$= \frac{y}{y + 5}$$

Recall:
$a^2 - b^2 = (a - b)(a + b)$

Note:
In practice, steps 2 and 3 would be done in one line.

To **add** or **subtract** algebraic fractions:

- Factorise the denominators.
- Express each term as a fraction whose denominator is the lowest common multiple of all the denominators (lowest common denominator).
- Add or subtract the fractions, writing a common denominator.
- Simplify the numerator, factorising if possible.
- Cancel common factors if possible.

Example 1.3 **a** Express $\dfrac{x}{x-4} - \dfrac{28}{x^2-x-12}$ as a fraction in its simplest form.

b Hence solve $\dfrac{x}{x-4} - \dfrac{28}{x^2-x-12} = 5$.

Step 1: Factorise the denominators.

a $\dfrac{x}{x-4} - \dfrac{28}{x^2-x-12} = \dfrac{x}{x-4} - \dfrac{28}{(x-4)(x+3)}$

Tip:
Form an equivalent fraction by multiplying the top and bottom of $\dfrac{x}{x-4}$ by $(x+3)$.

Step 2: Write each term as a fraction with the lowest common denominator.

$= \dfrac{x(x+3)}{(x-4)(x+3)} - \dfrac{28}{(x-4)(x+3)}$

Step 3: Add or subtract the fractions.

$= \dfrac{x(x+3) - 28}{(x-4)(x+3)}$

Step 4: Simplify the numerator.

$= \dfrac{x^2 + 3x - 28}{(x-4)(x+3)}$

$= \dfrac{(x-4)(x+7)}{(x-4)(x+3)}$

Tip:
Factorise the numerator if possible.

Step 5: Cancel common factors.

$= \dfrac{{}^1(x-4)(x+7)}{{}_1(x-4)(x+3)}$

$= \dfrac{x+7}{x+3}$

Step 1: Form an equation using the simplified form from **a**.

b $\dfrac{x}{x-4} - \dfrac{28}{x^2-x-12} = 5$

$\Rightarrow \qquad \dfrac{x+7}{x+3} = 5$

Tip:
Eliminate the denominator by multiplying both sides of the equation by $(x+3)$.

Step 2: Solve the equation.

$x + 7 = 5(x+3)$

$x + 7 = 5x + 15$

$4x = -8$

$x = -2$

Long division of polynomials

You can use the **algebraic division** methods that you met in *Core 2* with more complex divisors:

- Arrange the dividend and the divisor in order of descending powers. Include terms with a coefficient of zero.

- Divide the first term of the dividend by the first term of the divisor. This gives you the first term of the quotient.

- Multiply the whole divisor by this term and subtract the result from the dividend.

- Bring down as many terms as you need to make your new dividend.

- Repeat until you've used all the terms from your original dividend.

Note:
In the calculation $\frac{14}{5} = 2 \text{ r } 4 = 2\frac{4}{5}$, 14 is the *dividend*, 5 is the *divisor*, 2 is the *quotient* and 4 is the *remainder*.

Example 1.4 Divide $9x^3 - x + 5$ by $3x + 2$.

Step 1: $9x^3 \div 3x = 3x^2$, the first term of the quotient.

Step 2: Multiply $3x + 2$ by $3x^2$ and subtract from the dividend.

Step 3: $-6x^2 \div 3x = -2x$, the second term of the quotient.

Step 4: Multiply $3x + 2$ by $-2x$ and subtract from the dividend.

Step 5: $3x \div 3x = 1$, the third term of the quotient.

Step 6: Multiply $3x + 2$ by 1 and subtract from the dividend.

$$
\begin{array}{r}
3x^2 - 2x + 1 \\
3x + 2 \overline{\smash{)}9x^3 + 0x^2 - x + 5} \\
\underline{9x^3 + 6x^2 } \\
-6x^2 - x \\
\underline{-6x^2 - 4x } \\
3x + 5 \\
\underline{3x + 2} \\
3
\end{array}
$$

Tip:
Don't forget to include $0x^2$.

Step 7: Write out the answer.

So $\dfrac{9x^3 - x + 5}{3x + 2} = 3x^2 - 2x + 1$ remainder 3

i.e. $\dfrac{9x^3 - x + 5}{3x + 2} = 3x^2 - 2x + 1 + \dfrac{3}{3x + 2}$

Note:
$3x^2 - 2x + 1$ is the number of times $3x + 2$ divides into $9x^3 - x + 5$ completely. You put the remainder over $3x + 2$ because that is what you are dividing by.

Example 1.5 Divide $2x^3 - 1$ by $x^2 + 2$.

Step 1: $2x^3 \div x^2 = 2x$, the first term of the quotient.

Step 2: Multiply $x^2 + 2$ by $2x$ and subtract from the dividend.

$$
\begin{array}{r}
2x \\
x^2 + 0x + 2 \overline{\smash{)}2x^3 + 0x^2 + 0x - 1} \\
\underline{2x^3 + 0x^2 + 4x } \\
-4x - 1
\end{array}
$$

Step 3: Write out the answer.

$\dfrac{2x^3 - 1}{x^2 + 2} = 2x$ remainder $-4x - 1$

i.e. $\dfrac{2x^3 - 1}{x^2 + 2} = 2x + \dfrac{-4x - 1}{x^2 + 2} = 2x - \dfrac{4x + 1}{x^2 + 2}$

Tip:
Here you need $0x^2$ and $0x$.

Tip:
You know you've reached the remainder because the degree (highest power) of $-4x - 1$ is lower than the degree of $x^2 + 2$.

Note:
The fraction line acts as a bracket, so this is $2x - \dfrac{(4x + 1)}{x^2 + 2}$.

SKILLS CHECK **1A: Simplifying rational expressions**

1 Simplify fully:

a $\dfrac{x^2}{3x^2 - 2x}$

b $\dfrac{x^2 - 4}{x^2 + 3x - 10}$

c $\dfrac{4x + 12}{2x^2 - 18}$

d $\dfrac{3x^2 + 13x + 4}{2x^2 + 5x - 12}$

2 Express as a single fraction in its simplest form:

a $\dfrac{x^2 - 49}{x} \div \dfrac{x + 7}{x^2}$

b $\dfrac{10x + 15}{5x^2} \times \dfrac{x^2 + 4x}{2x + 3}$

c $\dfrac{x^3 - x}{4x^2} \times \dfrac{2x}{x + 1}$

d $\dfrac{4t^2 - 1}{2t^2 + 11t + 5} \div \dfrac{6t - 3}{t^2 + 5t}$

3 Express as a fraction in its simplest form:

a $\dfrac{5}{x-1} + \dfrac{3}{x+2}$

b $3 + \dfrac{2}{4-y}$

c $\dfrac{1}{x-2} - \dfrac{3}{x^2 - 7x + 10}$

d $\dfrac{2x}{x^2 - 2x - 3} + \dfrac{1}{x^2 - 1}$

 4 Divide $2x^4 - x^3 + 2x^2 - 5$ by $x^2 - 2$.

 5 a Simplify $\dfrac{x^3 + x^2 + x}{3x^2 - 2x} \times \dfrac{3x^2 + x - 2}{1 + x}$.

b Hence solve $\dfrac{x^3 + x^2 + x}{3x^2 - 2x} \times \dfrac{3x^2 + x - 2}{1 + x} = 3$.

SKILLS CHECK **1A EXTRA** is on the CD

1.2 Functions

Definition of a function. Domain and range of functions.

A **mapping** is a relationship between two sets of data. The set of elements being mapped is called the **domain** and the resulting set is called the **range**.

Consider the set {4, 3, 2, 1} being mapped to the set {8, 6, 4, 2}.

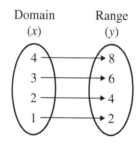

If the elements in the domain are represented by x and the elements in the range by y, the relationship is $y = 2x$.

Mappings can be described as:

- **one-to-one**   each element in the domain is mapped to just one element in the range;

- **many-to-one**   more than one element in the domain can be mapped to an element in the range;

- **one-to-many**   an element in the domain can be mapped to more than one element in the range;

- **many-to-many** more than one element in the domain can be mapped to more than one element in the range.

A **function** is a special mapping which satisfies both of the following conditions:

- It is defined for all elements of the domain.

- It is either one-to-one or many-to-one.

> **Note:**
> A mapping can be described in words or using algebra, and may be represented by a graph.

> **Note:**
> Mappings that are one-to-many or many-to-many are **not** functions.

Here are two examples of **one-to-one functions**, where each element in the domain maps to exactly one element in the range:

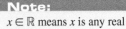

Note:
$x \in \mathbb{R}$ means x is any real number.

$$f : x \mapsto 3 - x, \ x \in \{3, 4, 5, 6\} \qquad f(x) = 2x - 1, \ x \in \mathbb{R}$$

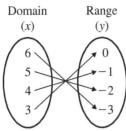

Range is $\{0, -1, -2, -3\}$

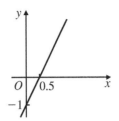

Range is $f(x) \in \mathbb{R}$

Here are two examples of **many-to-one functions**, where more than one element in the domain can map to an element in the range.

$$f(x) = x^4, \ x \in \{-2, -1, 0, 1, 2\} \qquad g : x \mapsto x^2 + 1, \ x \in \mathbb{R}$$

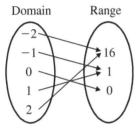

Range is $\{0, 1, 16\}$

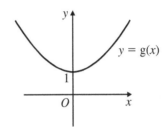

Range is $g(x) \in \mathbb{R}, \ g(x) \geqslant 1$

When a function is drawn as a graph it is easy to see the domain and range. The domain is the set of all possible x-values and the range is the set of all possible y-values.

Example 1.6 The diagrams show a sketch of the given mapping $x \mapsto y$. State, with a reason, whether or not the mapping is a function. If it is a function, state its range.

a $x^2 + y^2 = 25, \ x \in \mathbb{R}$ **b** $y = 4 - x^2, \ x \in \mathbb{R}$

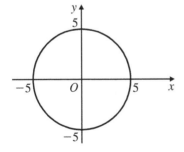

Step 1: Check the conditions for a mapping to be a function and make your conclusion.

a The mapping is many-to-many, since, for example, when $x = 3$, $y = \pm 4$ and also when $x = -3$, $y = \pm 4$.

Also the mapping is not defined for all elements of the domain, for example when $x = 6$, $y^2 = 25 - 36 = -9$, which has no real solutions. Hence $x^2 + y^2 = 25, \ x \in \mathbb{R}$ is not a function.

Recall:
Many-to-many mappings are not functions.

Step 2: If it is a function, use the sketch to state its range.

b The mapping is many-to-one and it is defined for all elements of the domain. So $y = 4 - x^2, \ x \in \mathbb{R}$ is a function.

The range is $y \in \mathbb{R}, \ y \leqslant 4$.

Tip:
Be careful with your use of $<$ and $\leqslant$. Here, when $x = 0$, $y = 4$, so the range is $y \leqslant 4$.

Example 1.7 The function f is such that $f(x) = x^2 - 4x - 5, x \in \mathbb{R}$.

a The diagram shows a sketch of $y = f(x)$.

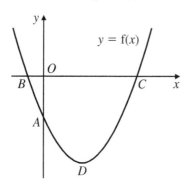

i Find the coordinates of A, B and C.

ii Write $f(x)$ in the form $(x - a)^2 + b$ and hence, or otherwise, find the coordinates of the minimum point D.

b State the range of f.

Step 1: Find the intercepts with the axes.

a i Consider $y = x^2 - 4x - 5$:

When $x = 0$, $y = -5$

When $y = 0$, $x^2 - 4x - 5 = 0$

$\Rightarrow \qquad (x - 5)(x + 1) = 0$

$\Rightarrow \qquad\qquad x = 5, x = -1$

Therefore A is the point $(0, -5)$, B is the point $(-1, 0)$ and C is the point $(5, 0)$.

Step 2: Complete the square to find a and b.

ii $f(x) = x^2 - 4x - 5$

$ = (x - 2)^2 - 4 - 5$

$ = (x - 2)^2 - 9$

Step 3: Write down the coordinates of the vertex of the curve.

So D has coordinates $(2, -9)$.

Step 4: Use the sketch and the coordinates of D to write down the range.

b Range of f is $f(x) \in \mathbb{R}, f(x) \geqslant -9$.

Example 1.8 $f(x) = \dfrac{3}{x - 2} - \dfrac{12}{x^2 - 4}, x > 2$

a Show that $f(x) = \dfrac{3}{x + 2}$.

b The diagram shows a sketch of $y = f(x)$.

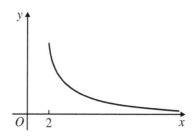

Find the range of f.

Step 1: Factorise the denominators.

a $\dfrac{3}{x-2} - \dfrac{12}{x^2-4} = \dfrac{3}{x-2} - \dfrac{12}{(x-2)(x+2)}$

Step 2: Write each term as a fraction with the lowest common denominator.

$= \dfrac{3(x+2)}{(x-2)(x+2)} - \dfrac{12}{(x-2)(x+2)}$

Step 3: Subtract the fractions.

$= \dfrac{3(x+2)-12}{(x-2)(x+2)}$

Step 4: Simplify the numerator.

$= \dfrac{3x+6-12}{(x-2)(x+2)}$

$= \dfrac{3x-6}{(x-2)(x+2)}$

Step 5: Cancel common factors.

$= \dfrac{3\cancel{(x-2)}^{1}}{_{1}\cancel{(x-2)}(x+2)}$

$= \dfrac{3}{(x+2)}$

Step 6: By considering the boundary values of the domain **and** using the sketch, state the range.

b $f(x) = \dfrac{3}{(x+2)}, x > 2$

When $x = 2$, $\dfrac{3}{(x+2)} = \dfrac{3}{(2+2)} = \dfrac{3}{4}$, but, since $x > 2$, $f(x) < \dfrac{3}{4}$.

From the graph, as $x \to \infty$, $f(x) \to 0$, so $f(x) > 0$

$\Rightarrow$ range of f is $f(x) \in \mathbb{R}$, $0 < f(x) < \frac{3}{4}$

SKILLS CHECK **1B: Definition, domain and range of a function**

1 State whether or not each of the following mappings represents a function. If it is a function, state whether it is one-to-one or many-to-one.

   a $y = 2(5-x), x \in \mathbb{R}$

   b $y^2 = 1 - x^2, x \in \mathbb{R}$

   c $y = \dfrac{1}{x}, x \in \mathbb{R}, x \neq 0$

   d $y = x^2, x \in \mathbb{R}$

2 Find the range of each of the following functions:

   a $f(x) = \dfrac{x+5}{2}, x = \{0, 1, 2, 3\}$

   b $f: x \mapsto \dfrac{1}{5-x}, x = \{1, 2, 3, 4\}$

   c $g: x \mapsto x^3, x \in \mathbb{R}, x \geqslant 0$

   d $f(x) = x^2 - 2, x \in \mathbb{R}$

3 For each of the following functions
   i sketch the function,   **ii** state its range.

   a $f: x \mapsto 4x - 3, x \in \mathbb{R}$

   ◎ **b** $g(x) = \sin x°, x \in \mathbb{R}, 0 \leqslant x \leqslant 360$

   c $f(x) = \dfrac{1}{x^2}, x \in \mathbb{R}, x \neq 0$

   d $h: x \mapsto x^2 - 6x, x \in \mathbb{R}$

◎ **4** The function f is defined by f: $x \mapsto \dfrac{6}{x} + 2x$ for the domain $1 \leqslant x \leqslant 3$.

Find the range of f.

SKILLS CHECK **1B EXTRA is on the CD**

1.3 Composite functions

Two or more functions may be combined to form a **composite function**.

The notation fg is used to denote the composite function f[g(x)].

For the composite function fg to exist, the range of g must be a subset of the domain of f.

> **Note:**
> fg means that you apply g then you apply f to the result.

Example 1.9 The functions f and g are defined as follows:

$$f: x \mapsto 2x + 3, x \in \mathbb{R} \qquad g: x \mapsto x^2 - 2, x \in \mathbb{R}$$

a Find **i** fg(x) **ii** gf(5) **iii** $f^2(x)$

b Solve gf(x) = −1.

Step 1: Apply g, then apply f to the result.
Step 2: Simplify.

a i $fg(x) = f(x^2 - 2)$
$= 2(x^2 - 2) + 3$
$= 2x^2 - 1$

> **Tip:**
> $fg(x) \neq f(x) \times g(x)$.

Step 3: Substitute the given value into f, then apply g to the result.
Step 4: Evaluate.

ii $gf(5) = g(2 \times 5 + 3)$
$= g(13)$
$= 13^2 - 2$
$= 167$

Step 5: Apply f, then apply f again to the result.
Step 6: Simplify.

iii $f^2(x) = f(2x + 3)$
$= 2(2x + 3) + 3$
$= 4x + 9$

> **Note:**
> $f^2(x) = ff(x)$.

> **Tip:**
> $f^2(x) \neq f(x) \times f(x)$.

Step 7: Apply f, then apply g to the result.

Step 8: Simplify.

b $gf(x) = g(2x + 3)$
$= (2x + 3)^2 - 2$
$= 4x^2 + 12x + 9 - 2$
$= 4x^2 + 12x + 7$

> **Note:**
> In general $fg(x) \neq gf(x)$.

Step 9: Set up an equation and solve for x.

$$gf(x) = -1$$
$$\Rightarrow 4x^2 + 12x + 7 = -1$$
$$4x^2 + 12x + 8 = 0$$
$$x^2 + 3x + 2 = 0$$
$$(x + 1)(x + 2) = 0$$
$$x = -1, x = -2$$

> **Note:**
> This doesn't mean find gf(−1).

> **Tip:**
> If the quadratic doesn't factorise, use the quadratic formula.

Example 1.10 The functions f and g are defined as follows:

$$f: x \mapsto x^2, x \in \mathbb{R} \qquad g: x \mapsto 3x - 2, x \in \mathbb{R}$$

a Find fg(x) and state its range.

b Find gf(x) and state its range.

c Find the value of a for which fg(a) = gf(a).

Step 1: Apply g, then apply f to the result.
Step 2: Simplify.

a $fg(x) = f(3x - 2)$
$= (3x - 2)^2$
$= 9x^2 - 12x + 4$

Step 3: State the range of fg.

Since $fg(x) = (3x - 2)^2$,
the range of fg is $fg(x) \in \mathbb{R}$, $fg(x) \geq 0$

> **Tip:**
> Use the fact that this is in completed square form so you know its minimum. (C1 Section 1.6). If you are unsure, do a quick sketch.

Step 4: Apply f, then apply g to the result.

Step 5: State the range of gf.

Step 6: Substitute the given variable, using your answers from **a** and **b**.

Step 7: Set up an equation and solve for a.

b $gf(x) = g(x^2)$
$$= 3x^2 - 2$$

Since $3x^2 \geqslant 0$ for all x, $3x^2 - 2 \geqslant -2$, so the range of gf is $gf(x) \in \mathbb{R}$, $gf(x) \geqslant -2$

c $fg(a) = 9a^2 - 12a + 4$
$gf(a) = 3a^2 - 2$
$fg(a) = gf(a)$
$\Rightarrow 9a^2 - 12a + 4 = 3a^2 - 2$
$6a^2 - 12a + 6 = 0$
$a^2 - 2a + 1 = 0$
$(a - 1)(a - 1) = 0$
$a = 1$

Tip:
Again, a quick sketch may help.

Tip:
You could substitute your answer into fg and gf to check that you get the same result.

1.4 Inverse functions

Inverse functions and their graphs.

The inverse function, f^{-1}, of a function f represents the reverse mapping. For a function to have an inverse, it must be one-to-one.
The domain of f^{-1} is the range of f.
The range of f^{-1} is the domain of f.

For example, consider $f : x \mapsto \frac{1}{2}x + 1$ with domain $\{3, 2, 1, 0\}$.

The range of f is $\{2.5, 2, 1.5, 1\}$.

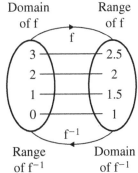

Now extend the domain so that $f : x \mapsto \frac{1}{2}x + 1$ for $x \in \mathbb{R}$. We can illustrate the function on a graph by the line $y = \frac{1}{2}x + 1$.

To show the inverse function, use the fact that the graphs of a function and its inverse are reflections of each other in the line $y = x$.

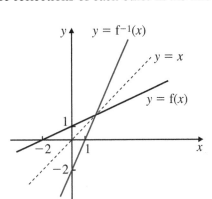

Note:
The notation $f^{-1}(x)$ is easy to confuse with $f'(x)$ and $(f(x))^{-1}$. $f^{-1}(x)$ is the inverse function, $f'(x)$ is the derivative and $(f(x))^{-1}$ is the reciprocal, i.e. $\dfrac{1}{f(x)}$.

Note:
The inverse function sends each element from the range of f back to its original value in the domain of f, so under f^{-1} $2.5 \mapsto 3, 2 \mapsto 2, 1.5 \mapsto 1$ and $1 \mapsto 0$.

Tip:
To show this on a diagram, ensure that the scales on both axes are the same.

Note:
In this example, the inverse function can be illustrated by a line.

So if (x, y) lies on f(x), then (y, x) lies on $f^{-1}(x)$.

To find the inverse function f^{-1}, let

$$y = \tfrac{1}{2}x + 1$$

Then interchange x and y:

$$x = \tfrac{1}{2}y + 1$$

Note:
You can interchange x and y at the beginning of the working or at the end.

Now make y the subject:

$$\tfrac{1}{2}y = x - 1$$
$$y = 2x - 2$$

Write this as a function:

$$f^{-1}: x \mapsto 2x - 2, \, x \in \mathbb{R}$$

Note:
This is the equation of the line $y = f^{-1}(x)$ shown in the diagram above.

Note:
Always write $f^{-1}(x)$ as a function of x.

If a function is many-to-one, its domain can be restricted to make it a one-to-one function. Then it will be possible to find the inverse function.

Consider the function $f(x) = x^2$, $x \in \mathbb{R}$. This is a many-to-one function since, for example, $f(-3) = 9$ and $f(3) = 9$.

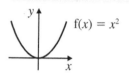

If the domain is restricted to $x \geq 0$, then f(x) becomes a one-to-one function and its inverse can be found.

Its inverse is $f^{-1}(x) = \sqrt{x}$, i.e. the positive square root of x.

Example 1.11 The function f has domain $3 \leq x \leq 9$ and is defined by

$$f(x) = \frac{8}{1 - x} + 6. \text{ A sketch of } y = f(x) \text{ is shown.}$$

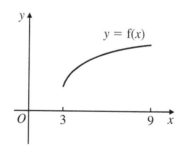

a Calculate f(3) and f(9).

b Find the range of f.

c The inverse function is f^{-1}. Find f^{-1}, stating its domain and range.

d On the same set of axes, sketch $y = f(x)$ and $y = f^{-1}(x)$.

Step 1: Substitute appropriate x values into the formula for f.

a $f(3) = \dfrac{8}{1 - 3} + 6 = 2$

$$f(9) = \dfrac{8}{1 - 9} + 6 = 5$$

Step 2: Use the values obtained in **a** and check the sketch.

b The range of f is $2 \leq f(x) \leq 5$.

Note:
You also need to check the sketch to make sure there are no turning points.

Step 3: Let $y = f(x)$ then interchange x and y.

c Let $y = \dfrac{8}{1 - x} + 6$.

Interchanging x and y gives

$$x = \frac{8}{1 - y} + 6$$

Step 4: Make y the subject.

$$x - 6 = \frac{8}{1 - y}$$

$$1 - y = \frac{8}{x - 6}$$

$$y = 1 - \frac{8}{x - 6}$$

$$y = \frac{x - 6 - 8}{x - 6}$$

$$y = \frac{x - 14}{x - 6}$$

Step 5: Use the relationship between f and f^{-1} to state the domain and range.

So $f^{-1}(x) = \dfrac{x - 14}{x - 6}$.

The domain of f^{-1} is $2 \leqslant x \leqslant 5$ and the range is $3 \leqslant f^{-1}(x) \leqslant 9$.

Step 6: Reflect the given curve in the line $y = x$ for the appropriate domain and range.

d

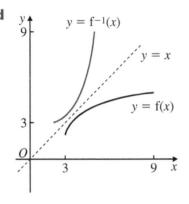

In the above example you can check your inverse function by substituting a value for x in the original function, for example:

$$f(3) = \frac{8}{1 - 3} + 6 = 2$$

Now substitute the result into the inverse function.

$$f^{-1}(2) = \frac{2 - 14}{2 - 6} = \frac{-12}{-4} = 3$$

Note:
f maps 3 to 2, and f^{-1} maps 2 back to 3.

So, because the inverse function reverses the process of the original function

$$f^{-1}f(x) = x$$

It is also true that

$$ff^{-1}(x) = x$$

This property can be used to solve equations.

11

Example 1.12 The function f(x) is defined by

$$f(x) = \sqrt{x + 2}, \; x \in \mathbb{R}, \; x \geqslant -2.$$

a Find the inverse function, $f^{-1}(x)$, stating its domain and range.

b Find the value of x for which $f(x) = \frac{1}{2}$.

c On the same set of axes, sketch $y = f(x)$ and $y = f^{-1}(x)$.

d Solve $f(x) = f^{-1}(x)$.

Step 1: Let $y = f(x)$ then interchange x and y.

a Let $y = \sqrt{x + 2}$

Interchanging x and y gives

$$x = \sqrt{y + 2}$$
$$x^2 = y + 2$$

Step 2: Make y the subject.

$$y = x^2 - 2$$
$$f^{-1}(x) = x^2 - 2$$

> **Tip:**
> Square both sides.

Step 3: Use the relationship between f and f^{-1} to state the domain and range.

Since the range of f is $f(x) \in \mathbb{R}$, $f(x) \geqslant 0$, the domain of f^{-1} is $x \in \mathbb{R}$, $x \geqslant 0$.

Since the domain of f is $x \in \mathbb{R}$, $x \geqslant -2$, the range of f^{-1} is $f^{-1}(x) \in \mathbb{R}$, $f^{-1}(x) \geqslant -2$.

> **Tip:**
> Make it clear for which function you are stating the domain or range.

Step 4: Use the inverse function as the reverse process.

b $f(x) = \frac{1}{2}$

$$\Rightarrow \quad x = f^{-1}\left(\tfrac{1}{2}\right)$$
$$= \left(\tfrac{1}{2}\right)^2 - 2$$
$$= -\tfrac{7}{4}$$

So $x = -\frac{7}{4}$.

> **Tip:**
> You can set up the equation $f(x) = \frac{1}{2} \Rightarrow \sqrt{x + 2} = \frac{1}{2}$ and solve it. That is an acceptable alternative method, but it may take longer.

Step 5: Sketch $f(x)$ for the domain and reflect it in $y = x$.

c

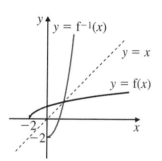

Step 6: Set up an equation in x and solve.

d From the graph, $f(x) = f^{-1}(x)$ at the point of intersection with the line $y = x$, i.e. when $f(x) = x$ and $f^{-1}(x) = x$.

Using $f^{-1}(x) = x$ gives

$$x^2 - 2 = x$$
$$x^2 - x - 2 = 0$$
$$(x - 2)(x + 1) = 0$$
$$x = 2 \text{ or } -1$$

But $x \geqslant 0$, so $x = 2$.

> **Tip:**
> It is much easier to solve $f(x) = x$ or $f^{-1}(x) = x$, rather than $f(x) = f^{-1}(x)$, for which you would have to solve $\sqrt{x + 2} = x^2 - 2$.
> Here $f^{-1}(x) = x$ is the easiest method as there are no $\sqrt{}$ signs to deal with.

> **Note:**
> From the sketch the intersection occurs when $x \geqslant 0$.

In the above example, the equation $f(x) = f^{-1}(x)$ is solved by finding any points at which $f(x) = x$ or $f^{-1}(x) = x$.

When $f(x) = x$ the elements are being mapped back onto themselves.

If $f(x) = x$ for *all* elements, the function is **self-inverse** and $f^{-1}(x)$ is the same as $f(x)$.

For example, the function $f(x) = \dfrac{1}{x}$, $x \neq 0$, is self-inverse because

$$f^{-1}(x) = \frac{1}{x}, x \neq 0.$$

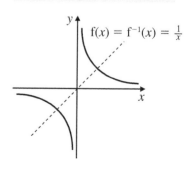

1 Functions f, g and h are defined as follows:

f: $x \mapsto 1 - 2x, x \in \mathbb{R}$ g: $x \mapsto x^2 + 3, x \in \mathbb{R}$ h: $x \mapsto \dfrac{x+5}{2}, x \in \mathbb{R}$

Find the value of

  a fg(2)   **b** hf(-1)   **c** gg(0)

  d fh(-11)   **e** gf($\frac{1}{4}$)   **f** hgf(1.5)

2 Functions f, g and h are defined as follows:

f: $x \mapsto 2^x, x \in \mathbb{R}$ g: $x \mapsto 3x + 2, x \in \mathbb{R}$ h: $x \mapsto \dfrac{1}{x}, x \in \mathbb{R}, x \neq 0$

Find the composite functions

  a fg   **b** hh   **c** gh   **d** hg

3 Functions f, g and h are defined as follows:

$f(x) = 2x + 9, x \in \mathbb{R}$ $g(x) = \log_{10} x, x \in \mathbb{R}, x > 0$ $h(x) = 1 - x^2, x \in \mathbb{R}$

Solve these equations:

  a ff$(x) = 9$ **b** gf$(x) = 0$ **c** fh$(x) = -5$ **d** hf$(x) = -8$

4 Which of the following functions has an inverse?

  a

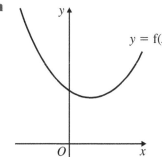

  b

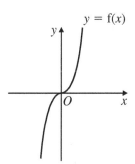

5 For each of the following functions, f,

  i find the inverse function f^{-1}, stating is domain and range,

  ii on the same set of axes, sketch $y = f(x)$ and $y = f^{-1}(x)$.

  a f: $x \mapsto 2x + 5, x \in \mathbb{R}$ **b** f: $x \mapsto \dfrac{3-x}{4}, x \in \mathbb{R}$

  c f: $x \mapsto x^2, x \in \mathbb{R}, x \geq 0$ **d** f: $x \mapsto \sqrt{x-3}, x \in \mathbb{R}, 3 \leq x \leq 12$

6 Functions f, g and h are defined as follows:

$f(x) = 5x - 4, x \in \mathbb{R}$ $g(x) = 1 - 2x, x \in \mathbb{R}$ $h(x) = x^2, x \in \mathbb{R}$

Solve these equations:

  a gf$(x) = g^{-1}(x)$ **b** h$(x) = g^{-1}(x)$ **c** hg$(x) = h(x)$

 7 A function is defined by f: $x \mapsto 3 - \dfrac{2}{x}, x \in \mathbb{R}, x \neq 0$.

  a Find f^{-1} and state the value of x for which f^{-1} is undefined.

  b Find the values of x for which $f(x) = f^{-1}(x)$.

8 A function is defined by f: $x \mapsto \dfrac{1}{5 - 4x}$, $x \in \mathbb{R}$, $x \neq \dfrac{5}{4}$.

    a Find the values of x which map onto themselves under the function f.

    b Find an expression for f^{-1}.

    Another function is defined by g: $x \mapsto x^2 - 3$.

    c Evaluate gf(1).

SKILLS CHECK **1C EXTRA** is on the CD

1.5 The modulus function

The modulus function.

The **modulus** (absolute value or magnitude) of a number is its positive numerical value. The modulus of x, $|x|$, is defined as

$$|x| = x, \qquad x \in \mathbb{R}, \quad x \geq 0$$
$$|x| = -x, \qquad x \in \mathbb{R}, \quad x < 0$$

For example, $|4| = |-4| = 4$.

> **Note:**
> The modulus is denoted by vertical lines as shown.

> **Tip:**
> Your calculator may have a modulus facility. It is often called 'abs'.

The graph of $y = |f(x)|$

When drawing the graph of $y = |f(x)|$, you need to consider the sign of $f(x)$.

- When $\mathbf{f(x) \geq 0}$, $|f(x)| = f(x)$, so the graph of $y = |f(x)|$ is the **same** as the graph of $y = f(x)$.

- When $\mathbf{f(x) < 0}$, $|f(x)| = -f(x)$, so the graph of $y = |f(x)|$ is a **reflection in the x-axis** of the graph of $y = f(x)$.

So, to draw $y = |f(x)|$ from the graph of $y = f(x)$, draw any part of $y = f(x)$ that is on or above the x-axis and replace any part that is below the x-axis with its reflection in the x-axis.

> **Tip:**
> When $f(x) > 0$, the graph of $y = f(x)$ is above the x-axis.

> **Tip:**
> When $f(x) < 0$, the graph of $y = f(x)$ is below the x-axis.

Example 1.13 f: $x \mapsto 3x - 2$, $x \in \mathbb{R}$

    a On separate sketches, draw $y = f(x)$ and $y = |f(x)|$.

    b The graph of $y = |f(x)|$ is made up of two lines. State the equations of these lines.

    c Find the values of x for which $|f(x)| = 4$.

Step 1: Draw a sketch of $y = 3x - 2$.

a

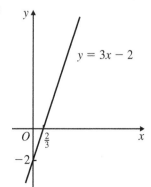

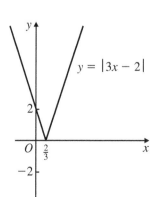

Step 2: Draw a second sketch, replacing the part of the line below the x-axis with its reflection in the x-axis.

> **Note:**
> $y = |3x - 2|$ consists of two parts: the line $y = 3x - 2$ when $3x - 2 \geq 0$ and the line $y = -(3x - 2)$ when $3x - 2 < 0$.

Step 3: State the equations of the lines.

b $y = 3x - 2$ and $y = -(3x - 2)$

Step 4: Set up two equations and solve.

c $3x - 2 = 4$ or $-(3x - 2) = 4$

$3x = 6$ $3x - 2 = -4$

$x = 2$ $3x = -2$

                                    $x = -\frac{2}{3}$

So $x = 2$ or $x = -\frac{2}{3}$

Tip:
To show this graphically, draw the line $y = 4$ and find the x-coordinates of the points of intersection of $y = |3x - 2|$ and $y = 4$.

Example 1.14 The diagram shows a sketch of $y = \sin x$, $0° \leqslant x \leqslant 360°$.

For $0° \leqslant x \leqslant 360°$,

a draw a sketch showing $y = |\sin x|$ and $y = 0.5$,

b solve the equation $|\sin x| = 0.5$.

Recall:
Trigonometric graphs
(C2 Section 4.5).

Step 1: Replace the curve below the x-axis with its reflection in the x-axis.

a

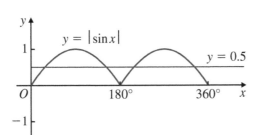

Step 2: Draw the given line on the same set of axes.

Tip:
Notice that two points of intersection are with the original part of the curve, i.e. $y = \sin x$, and the other two are with the reflected part of the curve, i.e. $y = -\sin x$.

Recall:
Find the principal value from the calculator and then find other solutions in range
(C2 Section 4.7).

Step 3: Solve the appropriate trig equations.

b $|\sin x| = 0.5$

$\Rightarrow$ $\sin x = 0.5$

        $x = 30°, 150°$

or $-\sin x = 0.5$

       $\sin x = -0.5$

          $x = 210°, 330°$

So $x = 30°, 150°, 210°, 330°$.

Example 1.15 **a** On the same set of axes sketch the graphs of $y = |2x - 6|$ and $y = \left|\dfrac{x}{2}\right|$.

b Solve the equation $|2x - 6| = \left|\dfrac{x}{2}\right|$.

Step 1: Draw a sketch of $y = |2x - 6|$.

a

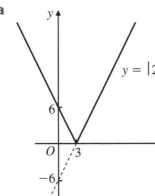

Tip:
Draw the graph of $y = 2x - 6$, but reflect the part below the x-axis in the x-axis.

Tip:
If you draw the line below the x-axis for guidance, make sure that it's clearly not part of your final answer.

Step 2: On the same diagram sketch $y = \left|\dfrac{x}{2}\right|$.

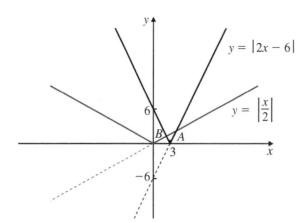

Tip:
The solutions for part **b** are the *x*-coordinates of the points of intersection, *A* and *B*.

Step 3: Set up two equations and solve.

b Intersection *A* is on the original part of both graphs, so

$$2x - 6 = \frac{x}{2}$$
$$4x - 12 = x$$
$$3x = 12$$
$$x = 4$$

Tip:
Use your sketch to determine which branch of the modulus to use when solving the equation.

Intersection *B* is on the reflected part of $y = 2x - 6$ i.e. $y = -(2x - 6)$ and the original part of $y = \dfrac{x}{2}$, so

$$-(2x - 6) = \frac{x}{2}$$
$$6 - 2x = \frac{x}{2}$$
$$12 - 4x = x$$
$$5x = 12$$
$$x = \frac{12}{5} = 2\tfrac{2}{5}$$

So $x = 4$ or $2\tfrac{2}{5}$.

Tip:
Take care with negatives and fractions.

The graph of $y = f(|x|)$

When drawing the graph of $y = f(|x|)$, you need to consider the sign of *x*.

- When $x \geqslant 0$, $f(|x|) = f(x)$, so the graph of $y = f(|x|)$ is the **same** as the graph of $y = f(x)$.

- When $x < 0$, $f(|x|) = f(-x)$, so the graph of $y = f(|x|)$ is a **reflection in the *y*-axis** of the graph of $y = f(x)$, $x > 0$.

So, to draw $y = f(|x|)$ from the graph of $y = f(x)$, disregard any part of $y = f(x)$ to the left of the *y*-axis. Draw a graph of the part of $y = f(x)$ to the right of the *y*-axis, then add its reflection in the *y*-axis.

Tip:
When $x > 0$, the graph of $y = f(x)$ is to the right of the *y*-axis.

Tip:
When $x < 0$, the graph of $y = f(x)$ is to the left of the *y*-axis.

Note:
$y = f(|x|)$ is symmetrical in the *y*-axis.

Example 1.16 The diagram shows a sketch of $y = f(x)$, where $f(x) = 3 + 2x - x^2$, $x \in \mathbb{R}$.

  a Sketch the graph of $y = f(|x|)$.

  b The graph of $y = f(|x|)$ and $y = -5$ intersect at *A* and *B*. Find the coordinates of *A* and *B*.

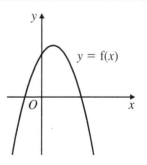

Step 1: Draw the given curve for $x > 0$, then reflect this in the y-axis.

a

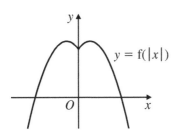

$y = f(|x|)$

Step 2: Draw $y = -5$ on the diagram.

Step 3: Find the positive solution of $f(x) = 0$ to find the coordinates of A.

b At A

$$3 + 2x - x^2 = -5$$
$$\Rightarrow \quad x^2 - 2x - 8 = 0$$
$$(x - 4)(x + 2) = 0$$
$$\Rightarrow \quad x = 4, x = -2 \text{ (not applicable)}$$

So A has coordinates $(4, -5)$.

Step 4: Use symmetry properties to find the coordinates of B.

Using symmetry, the coordinates of B are $(-4, -5)$.

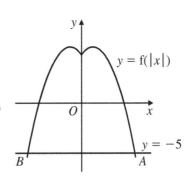

$y = f(|x|)$

$y = -5$

Example 1.17 $f(x) = x^2 - 6x, x \in \mathbb{R}$

The graph of $y = f(x)$ has a minimum point at C.

a Draw a sketch of $y = f(x)$, giving the coordinates of C and the coordinates of any intercepts with the axes.

b On separate sets of axes, sketch

i $y = |f(x)|$ **ii** $y = f(|x|)$

showing the coordinates of any intercepts with the axes and any stationary points.

c State the range of $|f(x)|$ and the range of $f(|x|)$.

Step 1: Substitute $y = 0$ and $x = 0$ to find the intercepts with the axes.

a Let $y = x^2 - 6x$

When $x = 0, y = 0$

When $y = 0, 0 = x^2 - 6x = x(x - 6)$

$x = 0$ or 6

Step 2: Complete the square to find the minimum point.

$y = (x - 3)^2 - 9$, so the minimum point is at $(3, -9)$.

Step 3: Sketch the curve.

C is the point $(3, -9)$ and the curve goes through $(0, 0)$ and $(6, 0)$.

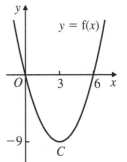

$y = f(x)$

Step 4: Sketch $y = |f(x)|$ by reflecting the part below the x-axis in the x-axis.

Step 5: Use symmetry to find the coordinates of the maximum point.

b i The graph meets the x-axis at $(0, 0)$ and $(6, 0)$.

By symmetry, the maximum point is $(3, 9)$.

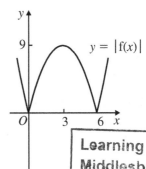

$y = |f(x)|$

Step 6: Sketch $y = \mathrm{f}(|x|)$ by reflecting the original graph to the right of the y-axis in the y-axis.

ii

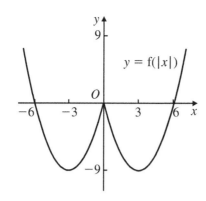

Step 7: Use symmetry to find the coordinates of the minimum point and the axes intercepts.

By symmetry, the graph meets the x-axis at $(-6, 0)$, $(0, 0)$ and $(6, 0)$.

The coordinates of the minimum points are $(-3, -9)$ and $(3, -9)$.

Step 8: Use your sketches to state the ranges.

c The range of $y = |\mathrm{f}(x)|$ is $y \in \mathbb{R}$, $y \geqslant 0$.

The range of $y = \mathrm{f}(|x|)$ is $y \in \mathbb{R}$, $y \geqslant -9$.

Recall:
The range is the set of possible values for y.

1.6 Combinations of transformations

Combinations of the transformations of $y = \mathrm{f}(x)$ represented by $y = a\mathrm{f}(x)$, $y = \mathrm{f}(x) + a$, $y = \mathrm{f}(x + a)$, $y = \mathrm{f}(ax)$.

Recall:
Transformations
(C1 Section 1.13).

In *Core 1* you learnt that transformations applied to $y = \mathrm{f}(x)$ have the following effects:

Translations
$y = \mathrm{f}(x) + a$ represents a translation by a units in the y-direction.
$y = \mathrm{f}(x + a)$ represents a translation by $-a$ units in the x-direction.

Tip:
The vector form of a translation by a units in the y-direction is $\begin{pmatrix} 0 \\ a \end{pmatrix}$.

Stretches
$y = a\mathrm{f}(x)$ represents a stretch by scale factor a in the y-direction.
$y = \mathrm{f}(ax)$ represents a stretch by scale factor $\frac{1}{a}$ in the x-direction.

Tip:
The vector form of a translation by $-a$ units in the x-direction is $\begin{pmatrix} -a \\ 0 \end{pmatrix}$.

Reflections
$y = -\mathrm{f}(x)$ represents a reflection in the x-axis.
$y = \mathrm{f}(-x)$ represents a reflection in the y-axis.

Combinations of transformations

In this unit you need to be able to apply combinations of these transformations. Here are some examples:

Note:
If, for example, $\mathrm{f}(x) = \sin x$, then $y = 2 \sin 5x$.

- $y = 2\mathrm{f}(5x)$ is a stretch of $y = \mathrm{f}(x)$ by scale factor $\frac{1}{5}$ in the x-direction and a stretch by scale factor 2 in the y-direction.

Note:
If, for example, $\mathrm{f}(x) = x^3$ then $y = -x^3 + 2$.

- $y = -\mathrm{f}(x) + 2$ is a reflection of $y = \mathrm{f}(x)$ in the x-axis followed by a translation of 2 units in the y-direction.

- $y = \mathrm{f}(x - 4) + 3$ is a translation of $y = \mathrm{f}(x)$ by 4 units in the x-direction and 3 units in the y-direction. The vector of this translation is $\begin{pmatrix} 4 \\ 3 \end{pmatrix}$.

Note:
If, for example, $y = x^2$ then $y = (x - 4)^2 + 3$. You used this in *Core 1* to sketch quadratic curves (C1 Section 1.13).

Sketching curves

Combinations of transformations can be used to sketch curves.

For example, consider the graph of $y = 2|x| + 3$. This is a stretch of $y = |x|$ by scale factor 2 in the y-direction to give $y = 2|x|$, followed by a translation of $y = 2|x|$ by 3 units in the y-direction.

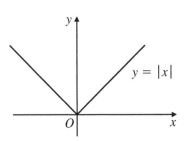

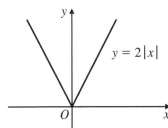

If $f(x) = |x|$, $y = 2f(x)$

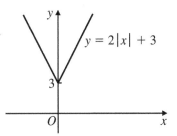

If $g(x) = 2|x|$, $y = g(x) + 3$

Example 1.18 The diagram shows a sketch of $y = f(x)$.

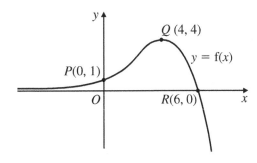

For each of the following, sketch the curve showing the coordinates of P_1, Q_1 and R_1, the images of P, Q and R.

a $y = 4 - f(x)$

b $y = \frac{1}{2}f(x + 3)$

c $y = \frac{1}{4}f(2x)$

Step 1: Apply the transformations to the given curve.

a This is a reflection of $y = f(x)$ in the x-axis, followed by a translation by 4 units in the y-direction.

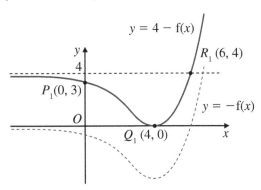

Step 2: Find the coordinates of the images of the given points.

The images of P, Q and R are $(0, 3)$, $(4, 0)$ and $(6, 4)$ respectively.

Tip:
The broken curve is the reflection. The solid curve is the final transformation. Make this clear on your sketch.

Note:
$y = 4$ is an asymptote of the curve $y = 4 - f(x)$.

Tip:
The reflection in the x-axis makes all the y-coordinates negative and then the vertical translation adds 4 to the y-coordinates. The x-values remain unchanged.

19

Step 1: Apply the transformations to the given curve.

b This is a translation of $y = f(x)$ by -3 units in the x-direction, and a stretch by scale factor $\frac{1}{2}$ in the y-direction.

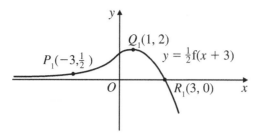

Tip:
The horizontal translation takes 3 off the x-coordinates and the vertical stretch halves the y-coordinates.

Step 2: Find the coordinates of the images of the given points.

The images of P, Q and R are $\left(-3, \frac{1}{2}\right)$, $(1, 2)$ and $(3, 0)$ respectively.

Step 1: Apply the transformations to the given curve.

c This is a stretch by scale factor $\frac{1}{2}$ in the x-direction, and a stretch by scale factor $\frac{1}{4}$ in the y-direction.

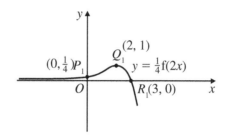

Tip:
The horizontal stretch halves the x-coordinates and the vertical stretch divides the y-coordinates by 4.

Step 2: Find the coordinates of the images of the given points.

The images of P, Q and R are $\left(0, \frac{1}{4}\right)$, $(2, 1)$ and $(3, 0)$ respectively.

Example 1.19 The diagram shows a sketch of the graph $y = f(x)$, where $f: x \mapsto \sin x$, $0 \leqslant x \leqslant 2\pi$.

Given that $g: x \mapsto 3 + \sin 2x$, $0 \leqslant x \leqslant 2\pi$,

a sketch the graph of $y = g(x)$,

b state the range of $g(x)$.

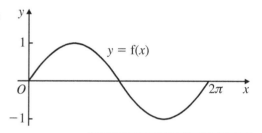

Step 1: Apply the transformations to the given curve.

a This is a stretch by scale factor $\frac{1}{2}$ in the x-direction, and a translation by 3 units in the y-direction.

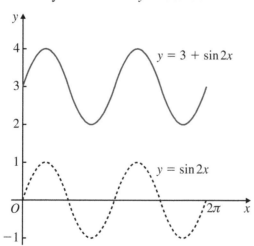

Tip:
If you draw intermediate steps, make it clear which graph is which.

Step 2: Use the sketch to find the range.

b From the graph, the range of $g(x)$ is $g(x) \in \mathbb{R}$, $2 \leqslant g(x) \leqslant 4$.

Tip:
The range of $\sin x$ is $-1 \leqslant y \leqslant 1$ so, following the translation by 3 units in the y-direction, this becomes $2 \leqslant y \leqslant 4$. Notice that the horizontal stretch doesn't affect the range.

1 Sketch the graphs of the following, giving the coordinates of any points of intersection with the coordinate axes.

   a $y = |2x - 3|$ **b** $y = |1 - 2x|$ **c** $y = |x| + 4$

2 For each of the following, on separate sets of axes sketch the graphs of $y = f(x)$ and $y = g(x)$, showing the coordinates of any points at which the curves have turning points or meet the axes.

   a $f(x) = x(x + 4)$ $g(x) = |x(x + 4)|$

   b $f(x) = \dfrac{1}{x}$ $g(x) = \dfrac{1}{|x|}$

   c $f(x) = \cos x, \, 0 \leqslant x \leqslant 2\pi$ $g(x) = |\cos x|, \, 0 \leqslant x \leqslant 2\pi$

3 Sketch the graph of each of the following functions, where a is a positive constant, giving the coordinates of any points of intersection with the coordinate axes.

   a $y = |x - a|$ **b** $y = |2x + a|$ **c** $y = |x + a| + a$

4 In each of the following:

   a On the same set of axes, sketch the graphs of $y = f(x)$ and $y = g(x)$.

   b Solve the equation $f(x) = g(x)$.

      i $f(x) = |3 - 2x|$ $g(x) = x + 1$

      ii $f(x) = |3x - 5|$ $g(x) = |2x + 1|$

    iii $f(x) = |x|$ $g(x) = 2|x - a|$, where a is a positive constant

5 Describe how the graph of $y = f(x)$ can be mapped to the graph of $y = g(x)$ by applying two transformations:

   a $f(x) = x$ $g(x) = \frac{1}{2}x - 3$

   b $f(x) = 2^x$ $g(x) = 0.4(2^{-x})$

  c $f(x) = \dfrac{1}{x}$ $g(x) = \dfrac{5}{x - 2}$

6 A sketch of the graph $y = f(x)$ is shown in the diagram. The point P has coordinates $(-2, 0)$, Q has coordinates $(0, 3)$ and R has coordinates $(2, 0)$.

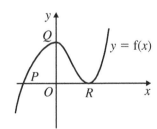

On separate sets of axes, sketch the graph of each of the following, showing the coordinates of any points at which the curve has a turning point.

   a $y = f(|x|) - 2$ **b** $y = -|f(x)|$ **c** $y = \frac{1}{3}f(2x)$ **d** $y = f(x - 3) + 2$

 7 **a** The graph of $y = \tan x$ is subjected to a stretch in the y-direction by scale factor 3 followed by a translation of -2 units in the y-direction. Write down the equation of the resulting curve.

    b The graph of $y = \sin x$ is subjected to a stretch in the x-direction by scale factor 3 followed by a reflection in the x-axis. Write down the equation of the resulting curve.

8 a Sketch the graph of $y = f(x)$, where f: $x \mapsto \cos x$, $0 \leqslant x \leqslant 2\pi$.

Given that g: $x \mapsto -\cos\left(x + \dfrac{\pi}{4}\right)$, $0 \leqslant x \leqslant 2\pi$,

b sketch the graph of $y = g(x)$, showing the coordinates of any turning points and intersections with the x-axis,

c state the range of $g(x)$.

SKILLS CHECK **1D EXTRA** is on the CD

Examination practice 1:   Algebra and functions

1 Express $\dfrac{3x^2 + 15x}{(2x + 1)^2} \div \dfrac{x^2 - 25}{2x^2 - 9x - 5}$ as a single fraction in its simplest form.

2 a 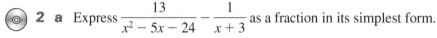 Express $\dfrac{13}{x^2 - 5x - 24} - \dfrac{1}{x + 3}$ as a fraction in its simplest form.

b Hence solve $\dfrac{13}{x^2 - 5x - 24} - \dfrac{1}{x + 3} = 1$.

3 The functions f and g are defined by

$$\text{f: } x \mapsto 2x + 1, x \in \mathbb{R},$$

$$\text{g: } x \mapsto \frac{1}{x}, \qquad x \in \mathbb{R}, x \neq 0.$$

a Calculate the value of gf(2).

b Find g^{-1}.

c Calculate the values of x for which fg(x) = x. [London June 1992]

4 The functions f and g are defined for all real values of x by

$$\text{f: } x \mapsto x^2,$$
$$\text{g: } x \mapsto 4 - 9x.$$

a Express the composite function gf in terms of x.

b Sketch the curve with equation $y = gf(x)$ and show on your sketch the coordinates of the points at which your curve intersects the x-axis.

c Determine the range of the function gf.

d Find the value of x for which g(x) = $g^{-1}(x)$, where g^{-1} is the inverse function of g. [London Jan 1996]

5 $f(x) = \dfrac{2}{x - 1} - \dfrac{6}{(x - 1)(2x + 1)}$, $x > 1$.

a Prove that $f(x) = \dfrac{4}{2x + 1}$.

b Find the range of f.

c Find $f^{-1}(x)$.

d Find the range of $f^{-1}(x)$. [Edexcel Jan 2002]

6 The functions f and g are defined by

$$f: x \mapsto 4x - 1,\ x \in \mathbb{R},$$

$$g: x \mapsto \frac{3}{2x - 1},\ x \in \mathbb{R},\ x \neq \tfrac{1}{2}.$$

Find in its simplest form

a the inverse function f^{-1},

b the composite function gf, stating its domain.

c Find the values of x for which

$$2f(x) = g(x),$$

giving your answers to 3 decimal places.

[London June 1997]

7 Functions f and g are defined by

$$f: x \mapsto 4 - x,\ x \in \mathbb{R},$$
$$g: x \mapsto 3x^2,\ x \in \mathbb{R},$$

a Find the range of g.

b Solve $gf(x) = 48$.

c Sketch the graph of $y = |f(x)|$ and hence find the values of x for which $|f(x)| = 2$.

[London Jan 1996]

8 a On the same axes, sketch the graphs of $y = 2 - x$ and $y = 2|x + 1|$.

b Hence, or otherwise, find the values of x for which $2 - x = 2|x + 1|$.

[London Jan 1997]

9 The functions f and g are defined as follows:

$$f: x \mapsto a - x,\ x \in \mathbb{R} \qquad g: x \mapsto |2x + a|,\ x \in \mathbb{R},$$

where a is a positive constant.

a Find $gf(4a)$.

b On the same set of axes, sketch the graphs of $y = f(x)$ and $y = g(x)$.

c Solve the equation $f(x) = g(x)$.

10

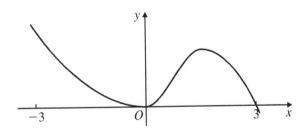

The diagram shows a sketch of the curve with equation $y = f(x)$.

In separate diagrams show, for $-3 \leqslant x \leqslant 3$, sketches of the curves with equation

a $y = f(-x)$,

b $y = -f(x)$,

c $y = f(|x|)$.

Mark on each sketch the x-coordinate of any point, or points, where a curve touches or crosses the x-axis.

[London June 1993]

11

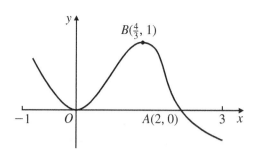

The diagram shows a sketch of the curve with equation $y = f(x)$, $-1 \leqslant x \leqslant 3$. The curve touches the x-axis at the origin O, crosses the x-axis at the point $A(2, 0)$ and has a maximum at the point $B\left(\frac{4}{3}, 1\right)$.

In separate diagrams, show a sketch of the curve with equation

a $y = f(x + 1)$, **b** $y = |f(x)|$, **c** $y = f(|x|)$,

marking on each sketch the coordinates of points at which the curve

    i   has a turning point,

    ii meets the x-axis.

[Edexcel June 2003]

 12

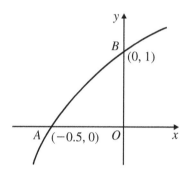

The diagram shows a sketch of the curve with equation $y = f(x)$. The curve crosses the x-axis at $(-0.5, 0)$ and the y-axis at $(0, 1)$.

Give the coordinates of any points of intersection with the coordinate axes of the following curves.

a $y = f(|x|) + 1$ **b** $y = 2f(-x)$ **c** $y = f^{-1}(x)$

13 The diagram shows part of the graph of $y = f(x)$, $x \in \mathbb{R}$. The graph consists of two line segments that meet at the point $(1, a)$, $a < 0$. One line meets the x-axis at $(3, 0)$. The other line meets the x-axis at $(-1, 0)$ and the y-axis at $(0, b)$, $b < 0$.

In separate diagrams, sketch the graph with equation

a $y = f(x + 1)$,

b $y = f(|x|)$.

Indicate clearly on each sketch the coordinates of any points of intersection with the axes.

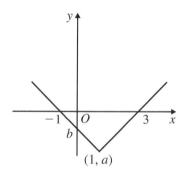

Given that $f(x) = |x - 1| - 2$, find

c the value of a and the value of b,

d the value of x for which $f(x) = 5x$.

[Edexcel June 2005]

2 Trigonometry

2.1 Inverse trigonometric functions

> Knowledge of arcsin, arccos and arctan. Their relationship to sine, cosine and tangent. Understanding of their graphs and appropriate restricted domains.

The three trigonometric functions of sine, cosine and tangent are many-to-one functions, since, for example,

$$\sin 30° = \sin 150° = \sin 390° = \ldots = 0.5$$

Since they are many-to-one functions, they do not have an inverse.

However, if we restrict the domain it is possible to define the **inverse trigonometric functions**: arcsin x, arccos x and arctan x, sometimes written $\sin^{-1} x$, $\cos^{-1} x$ and $\tan^{-1} x$.

The graphs of the inverse functions are shown below. They have been drawn using the following properties of functions.
For a function f and its inverse function f^{-1}:

- the domain of f^{-1} is the range of f
- the range of f^{-1} is the domain of f
- the graph of $y = f^{-1}(x)$ is a reflection in the line $y = x$ of the graph of $y = f(x)$.

> **Recall:**
> Only one-to-one functions have an inverse (Section 1.2).

> **Note:**
> You can use degrees or radians.

> **Note:**
> To show the reflection, the scales on both axes must be the same, so radians are used in the diagrams below.

arcsin x

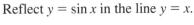

f: $x \mapsto \sin x$	Reflect $y = \sin x$ in the line $y = x$.	f^{-1}: $x \mapsto \arcsin x$ (or $\sin^{-1} x$)
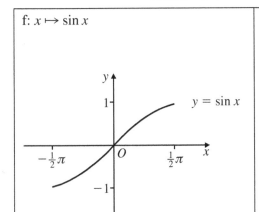	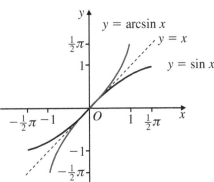	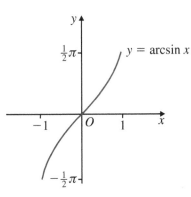
Domain $-\frac{1}{2}\pi \leqslant x \leqslant \frac{1}{2}\pi$		Domain $-1 \leqslant x \leqslant 1$
Range $-1 \leqslant \sin x \leqslant 1$		Range $-\frac{1}{2}\pi \leqslant \arcsin x \leqslant \frac{1}{2}\pi$

Notice that if you turn your page through a quarter turn clockwise, and imagine the axis that is now horizontal as the x-axis, you can see a reflection in that axis of the sine curve.

arccos x

f: $x \mapsto \cos x$	Reflect $y = \cos x$ in the line $y = x$.	f^{-1}: $x \mapsto \arccos x$ (or $\cos^{-1} x$)
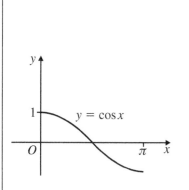	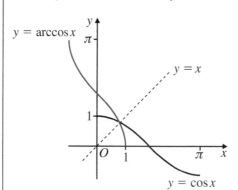	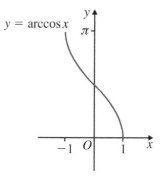
Domain $0 \leqslant x \leqslant \pi$		Domain $-1 \leqslant x \leqslant 1$
Range $-1 \leqslant \cos x \leqslant 1$		Range $0 \leqslant \arccos x \leqslant \pi$

Notice that if you turn your page through a quarter turn clockwise, and imagine the axis that is now horizontal as the x-axis, you can see a reflection in that axis of the cos curve.

arctan x

f: $x \mapsto \tan x$	Reflect $y = \tan x$ in the line $y = x$.	f^{-1}: $x \mapsto \arctan x$ (or $\tan^{-1} x$)
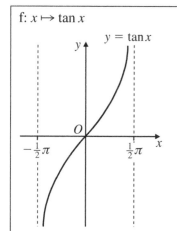	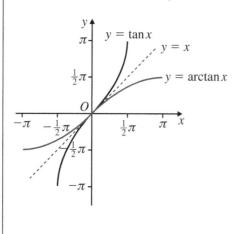	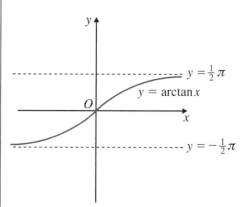
Domain $-\frac{1}{2}\pi \leqslant x \leqslant \frac{1}{2}\pi$		Domain $x \in \mathbb{R}$
Range $\tan x \in \mathbb{R}$		Range $-\frac{1}{2}\pi \leqslant \arctan x \leqslant \frac{1}{2}\pi$

The lines $y = -\frac{1}{2}\pi$ and $y = \frac{1}{2}\pi$ are asymptotes to the curve $y = \arctan x$.

Notice that if you turn your page through a quarter turn clockwise, and imagine the axis that is now horizontal as the x-axis, you can see a reflection in that axis of the tan curve.

Principal value

Remember that the value of $\arcsin x$, $\arccos x$ or $\arctan x$ is an **angle**. It is the angle given on a calculator by the inverse trig functions, labelled $\sin^{-1}$, $\cos^{-1}$ and $\tan^{-1}$, and is often referred to as the **principal value** (PV).

> **Recall:**
> The PV is used when solving trig equations. You can use degrees or radians (C2 Section 4.7).

Example 2.1 **a** Use a calculator to find, in degrees, the value of:

   i   arcsin 1   **ii**   arccos 0.5

   iii   arctan(−1)   **iv**   arccos(−0.3)

b Use a calculator to find, in radians correct to two decimal places,
the value of:

   i   arcsin(−0.6)   **ii**   arccos 0.3   **iii**   arctan 10

Step 1: Use appropriate
inverse trig functions on
the calculator.
a **i** $\arcsin 1 = 90°$ **ii** $\arccos 0.5 = 60°$

   iii $\arctan(-1) = -45°$ **iv** $\arccos(-0.3) = 107.45\ldots°$

b **i** $\arcsin(-0.6) = -0.6435\ldots = -0.64^c$ (2 d.p.)

   ii $\arccos 0.3 = 1.2661\ldots = 1.27^c$ (2 d.p.)

   iii $\arctan 10 = 1.4711\ldots = 1.47^c$ (2 d.p.)

> **Tip:**
> In part **a**, set your calculator to degree mode. In part **b**, set your calculator to radian mode.

Special angles

It is useful to learn the values of sin, cos and tan of these special
angles. They are especially useful if you are asked to give *exact*
values, in terms of π, for arcsin x, arccos x or arctan x.

> **Recall:**
> Special angles (C2 Section 4.5).

For example,

$$\arcsin \frac{1}{\sqrt{2}} = \frac{1}{4}\pi$$

$$\arccos\left(-\frac{1}{2}\right) = \frac{2}{3}\pi$$

$$\arctan \frac{1}{\sqrt{3}} = \frac{1}{6}\pi$$

$x°$	x^c	$\sin x$	$\cos x$	$\tan x$
30°	$\frac{1}{6}\pi$	$\frac{1}{2}$	$\frac{\sqrt{3}}{2}$	$\frac{1}{\sqrt{3}}$
45°	$\frac{1}{4}\pi$	$\frac{1}{\sqrt{2}}$	$\frac{1}{\sqrt{2}}$	1
60°	$\frac{1}{3}\pi$	$\frac{\sqrt{3}}{2}$	$\frac{1}{2}$	$\sqrt{3}$

Calculator note

When finding arcsin, arccos or arctan, the value given by the
calculator in radians is a decimal. To write it as a multiple of π, divide
the decimal value by π, then check whether the multiple can be
written as a fraction by pressing the fraction key. You must beware,
however! This method will probably not work if you have already
rounded values prior to this in a calculation. It is better to learn and be
able to recognise the trigonometric ratios of the special angles.

> **Tip:**
> This is useful if you forget the values of sin, cos and tan of the special angles, but do not rely on it as a method.

2.2 Reciprocal trigonometric functions

Knowledge of secant, cosecant and cotangent and their relationship to sine, cosine and
tangent. Understanding of their graphs. Knowledge and use of $\sec^2 \theta = 1 + \tan^2 \theta$ and
$\csc^2 \theta = 1 + \cot^2 \theta$.

The **reciprocal functions** of the three main trig functions of sine (sin),
cosine (cos) and tangent (tan) are **cosecant** (cosec), **secant** (sec) and
cotangent (cot). They are defined as follows:

> **Note:**
> Do not confuse the notation for the *reciprocal* function with the *inverse* function. For example, cosec x can be written $(\sin x)^{-1}$, whereas the inverse sine function arcsin x can be written $\sin^{-1} x$.

- $\csc x = \dfrac{1}{\sin x}$

- $\sec x = \dfrac{1}{\cos x}$

- $\cot x = \dfrac{1}{\tan x}$

Since $\tan x = \dfrac{\sin x}{\cos x}$, we can also write $\cot x = \dfrac{\cos x}{\sin x}$.

Graphs of reciprocal functions

You should learn the properties of the graphs of $y = \operatorname{cosec} x$, $y = \sec x$ and $y = \cot x$. It is useful to understand how they are obtained from the graphs of $y = \sin x$, $y = \cos x$ and $y = \tan x$ and these are also shown in the diagrams below.

Note:

A graphical calculator can be used in C3.

$y = \operatorname{cosec} x$

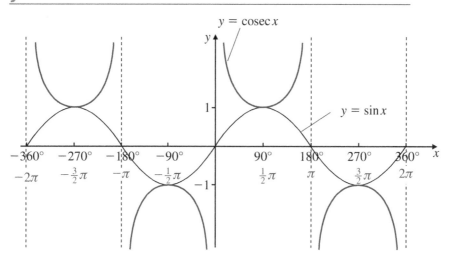

- Whereas $\sin x$ takes values between -1 and 1, $\operatorname{cosec} x$ is always greater than or equal to 1, or less than or equal to -1.
- There are vertical asymptotes through the points where $y = \sin x$ crosses the x-axis.
- There are minimum points where $y = \sin x$ has maximum points.
- There are maximum points where $y = \sin x$ has minimum points.

$y = \sec x$

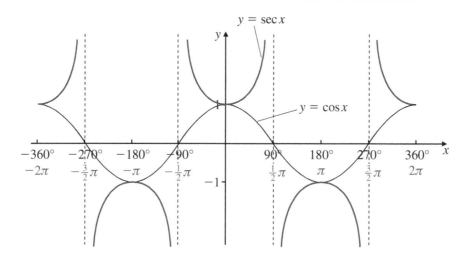

- The value of $\sec x$ is always greater than or equal to 1, or less than or equal to -1.
- There are vertical asymptotes through the points where $y = \cos x$ crosses the x-axis.
- There are minimum points where $y = \cos x$ has maximum points.
- There are maximum points where $y = \cos x$ has minimum points.

$y = \cot x$

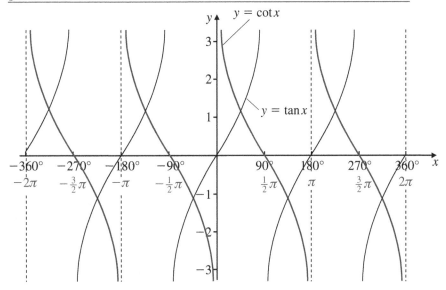

- $\cot x$ can take all values.

- There are vertical asymptotes through the points where $y = \tan x$ crosses the x-axis.

- The graph of $y = \cot x$ crosses the x-axis where $y = \tan x$ has vertical asymptotes.

Applying transformations

You could be asked to apply combinations of transformations to the graphs of $y = \operatorname{cosec} x$, $y = \sec x$ and $y = \cot x$.

Recall:
Transformations (Section 1.6).

Example 2.2 $f(x) = 2 \sec x + 1$.

a The graph of $y = f(x)$ can be obtained from the graph of $y = \sec x$ by applying a stretch followed by a translation.

   i State the scale factor and direction of the stretch.

   ii Describe the translation.

   iii State the coordinates of the image of the point $(0, 1)$ under this mapping.

b Write down the range of f for $-90° < x < 90°$.

Step 1: Compare with $y = af(x) + b$ and describe the stretch and the translation.

a i The stretch is in the y-direction and the scale factor is 2.

   ii The translation is by 1 unit in the y-direction.

Note:
The vector of the translation is $\begin{pmatrix} 0 \\ 1 \end{pmatrix}$.

Step 2: Apply the transformations in turn.

   iii Under the stretch, the image of $(0, 1)$ is $(0, 2)$.
   Under the translation, the image of $(0, 2)$ is $(0, 3)$.
   So the image of $(0, 1)$ is $(0, 3)$.

Tip:
For $-90° < x < 90°$, the minimum point of $y = \sec x$ is $(0, 1)$ and the range is $\sec x \geqslant 1$. What is the minimum point of $y = 2 \sec x + 1$?

Step 3: Identify the range of f from the information in **a**.

b The range of f for $-90° < x < 90°$ is $f(x) \geqslant 3$.

Identities involving reciprocal functions

In *Core 2* you used this identity:
$$\cos^2 \theta + \sin^2 \theta \equiv 1 \qquad \textcircled{1}$$

Recall:
Trig identities (C2 Section 4.6).

Manipulating this identity gives two further identities which must be learnt for this unit:
$$1 + \tan^2 \theta \equiv \sec^2 \theta \qquad \textcircled{2}$$
$$\cot^2 \theta + 1 \equiv \operatorname{cosec}^2 \theta \qquad \textcircled{3}$$

Tip:
Divide each term of identity ① by $\cos^2 \theta$.

Remember that these are identities, so they are true for all values of θ in degrees or in radians.

Tip:
Divide each term of identity ① by $\sin^2 \theta$.

You may be asked to use these identities to prove further identities or solve equations, as in the following examples.

Example 2.3 Solve the equation
$$\sec^2 \theta = 5(\tan \theta - 1),$$
where $0° \leqslant \theta \leqslant 360°$, giving your answers in degrees, correct to the nearest degree.

Tip:
Set your calculator to degrees mode.

Step 1: Use an appropriate identity to form an equation in $\tan \theta$.
$$\sec^2 \theta = 5(\tan \theta - 1)$$
$$\Rightarrow \quad 1 + \tan^2 \theta = 5 \tan \theta - 5$$
$$\Rightarrow \quad \tan^2 \theta - 5 \tan \theta + 6 = 0$$

Tip:
The linear term is $\tan \theta$, so write $\sec^2 \theta$ in terms of $\tan \theta$.

Step 2: Solve the equation in $\tan \theta$.
$$(\tan \theta - 3)(\tan \theta - 2) = 0$$
$$\Rightarrow \quad \tan \theta - 3 = 0$$
$$\tan \theta = 3$$
$$\theta = 71.5\ldots°,$$
$$251.5\ldots°$$

or
$$\tan \theta - 2 = 0$$
$$\tan \theta = 2$$
$$\theta = 63.4\ldots°,$$
$$243.4\ldots°$$

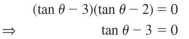

Tip:
This is a quadratic equation in $\tan \theta$.

Tip:
Since the tan function repeats every 180°, the values of θ in range are PV and PV + 180° (C2 Section 4.5).

So, to the nearest degree $\theta = 63°, 72°, 243°, 252°$.

Example 2.4 Find the exact values of θ, where $-\pi \leqslant \theta \leqslant \pi$, such that
$$2 \cot^2 \theta = 3 \operatorname{cosec} \theta$$

Tip:
Notice the description *exact*.

Step 1: Use an appropriate identity to form an equation in $\operatorname{cosec} \theta$.
$$2 \cot^2 \theta = 3 \operatorname{cosec} \theta$$
$$2(\operatorname{cosec}^2 \theta - 1) = 3 \operatorname{cosec} \theta$$
$$2 \operatorname{cosec}^2 \theta - 2 = 3 \operatorname{cosec} \theta$$
$$2 \operatorname{cosec}^2 \theta - 3 \operatorname{cosec} \theta - 2 = 0$$

Tip:
The linear term is $\operatorname{cosec} \theta$, so aim to form an equation just in $\operatorname{cosec} \theta$ using the relationship $1 + \cot^2 \theta \equiv \operatorname{cosec}^2 \theta$.

Step 2: Factorise and solve.
$$(2 \operatorname{cosec} \theta + 1)(\operatorname{cosec} \theta - 2) = 0$$
$$\Rightarrow \quad 2 \operatorname{cosec} \theta + 1 = 0$$
$$2 \operatorname{cosec} \theta = -1$$
$$\operatorname{cosec} \theta = -\tfrac{1}{2} \text{ (no solutions)}$$

or
$$\operatorname{cosec} \theta - 2 = 0$$
$$\operatorname{cosec} \theta = 2$$
$$\sin \theta = \tfrac{1}{2}$$
$$\theta = \tfrac{1}{6}\pi$$

or
$$\theta = \pi - \tfrac{1}{6}\pi = \tfrac{5}{6}\pi$$

So
$$\theta = \tfrac{1}{6}\pi, \tfrac{5}{6}\pi$$

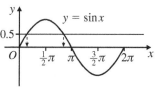

Tip:
Exact values are requested, so look out for special angles. You must not give an approximate answer.

Example 2.5 Prove the identity

$$(\sec x - \operatorname{cosec} x)(\sec x + \operatorname{cosec} x) \equiv (\tan x - \cot x)(\tan x + \cot x).$$

Note:
In a proof, all working must be shown.

General steps for proving an identity:

Step 1: Start with one side of the identity and simplify/use appropriate identities to write it in a different format.

Step 2: Continue rewriting/simplifying until you get the expression on the other side of the identity.

$$\begin{aligned}
\text{LHS} &= (\sec x - \operatorname{cosec} x)(\sec x + \operatorname{cosec} x) \\
&= \sec^2 x - \operatorname{cosec}^2 x \\
&= (1 + \tan^2 x) - (1 + \cot^2 x) \\
&= 1 + \tan^2 x - 1 - \cot^2 x \\
&= \tan^2 x - \cot^2 x \\
&= (\tan x - \cot x)(\tan x + \cot x) \\
&= \text{RHS}
\end{aligned}$$

So $(\sec x - \operatorname{cosec} x)(\sec x + \operatorname{cosec} x) \equiv (\tan x - \cot x)(\tan x + \cot x)$.

Recall:
Factorising the difference of two squares, where $(a - b)(a + b) = a^2 - b^2$.

When proving an identity, you can start with either side. So you can start with the left-hand side and aim to get to the expression on the right-hand side, or vice versa. This gives a neat method of proof, but it is also acceptable to show that each side is equal to the same (third) expression. This is sometimes referred to as 'meeting in the middle'.

SKILLS CHECK **2A: Trigonometric functions, inverses and identities**

1 Use your calculator to find the value, to the nearest degree, of each of the following:

   a $\arccos 0.45$ **b** $\arcsin(-0.67)$ **c** $\arctan 2.8$ **d** $\arccos(-0.2)$

2 Find the exact value, in radians in terms of π, of each of the following:

   a $\arctan 1$ **b** $\arcsin(-0.5)$ **c** $\arccos 0$ **d** $\arccos\left(\frac{1}{\sqrt{2}}\right)$

3 Solve the following equations, where $0° < x < 360°$. If your answer is not exact, give it to the nearest $0.1°$.

   a $\operatorname{cosec} x = 2.5$ **b** $\cot x = -\sqrt{3}$ **c** $\sec 2x = 1.5$ **d** $\sec^2 x = 4$

4 Solve the following equations, where $0 \leqslant \theta \leqslant 2\pi$. If your answer is not exact, give it to three significant figures.

   a $\cot \frac{1}{2}\theta = 0.4$ **b** $\operatorname{cosec} \theta = -1$ **c** $\sec(\theta + \frac{1}{2}\pi) = 4$ **d** $\operatorname{cosec}^2 \theta = 2$

5 Prove these identities:

   a $\tan \theta + \cot \theta \equiv \sec \theta \operatorname{cosec} \theta$ **b** $\sec \theta - \cos \theta \equiv \tan \theta \sin \theta$

6 Solve $\tan^2 \theta = 1 + \sec \theta$ for $0° \leqslant \theta \leqslant 360°$.

7 Solve $4 \operatorname{cosec} \theta - 5 = \cot^2 \theta$ for $-360° \leqslant \theta \leqslant 360°$.

8 Solve $\cot \theta = 2 \cos \theta$ for $0 < \theta \leqslant 2\pi$, giving your answers exactly in terms of π.

9 Find the exact values of θ, where $-\pi \leqslant \theta \leqslant \pi$, such that $\operatorname{cosec}^2 \theta = 2 \cot \theta$.

10 $f(x) = -\cot(x + 90°)$.

   a Describe how the graph of $y = f(x)$ may be obtained from the graph of $y = \cot x$.

   b On the same set of axes, sketch the graph of $y = \cot x$ and $y = -\cot(x + 90°)$ for $-180° \leqslant x \leqslant 180°$, labelling each curve clearly.

   c Describe the relationship between $y = -\cot(x + 90°)$ and $y = \tan x$.

11 $f(x) = 2 \operatorname{cosec} x + 1$

    a Describe transformations that map the graph of $y = \operatorname{cosec} x$ onto the graph of $y = f(x)$.

    b State the coordinates of the image of the point $(\frac{1}{2}\pi, 1)$ under this mapping.

    c Write down the range of f for $0 < x < \pi$.

12 **a** Describe transformations that map the graph of $y = \sec x°$ onto the graph of $y = \sec 2x° + 1$.

    b State the image of $(360, 1)$ under the mapping.

SKILLS CHECK **2A EXTRA is on the CD**

2.3 Addition and double angle formulae

Knowledge and use of the double angle formulae; use of formulae for
$\sin(A \pm B)$, $\cos(A \pm B)$ and $\tan(A \pm B)$.

Addition formulae

Note:
Despite their name, the formulae relate to two angles added or subtracted, not just added. They are also called compound angle formulae.

The addition formulae give expressions for sin, cos and tan of the **sum** or **difference** of two angles, A and B. They are identities and so are true for all values of A and B.

The six formulae are summarised in your formulae booklet as follows:

$$\sin(A \pm B) = \sin A \cos B \pm \cos A \sin B$$

$$\cos(A \pm B) = \cos A \cos B \mp \sin A \sin B$$

$$\tan(A \pm B) = \frac{\tan A \pm \tan B}{1 \mp \tan A \tan B}$$

Tip:
Make sure that you understand how to use $\pm$ and $\mp$ symbols, for example $\cos(A - B)$
$= \cos A \cos B + \sin A \sin B$.

Example 2.6 **a** Given that $\cos(\theta + 30°) = \sin\theta$, show that $\tan\theta = \frac{1}{\sqrt{3}}$.

    b Hence find all the values of θ, where $0° < \theta < 360°$, for which $\cos(\theta + 30°) = \sin\theta$.

Step 1: Expand the sum using the addition formula.

a
$$\cos(\theta + 30°) = \sin\theta$$
$$\Rightarrow \quad \cos\theta\cos 30° - \sin\theta\sin 30° = \sin\theta$$

Step 2: Insert known trig values.

$$\Rightarrow \quad \cos\theta \times \frac{\sqrt{3}}{2} - \sin\theta \times \frac{1}{2} = \sin\theta$$

Recall:
Trigonometric ratios of special angles (Section 2.1).

Step 3: Simplify to the required form.

Multiplying all the terms by 2 gives
$$\sqrt{3}\cos\theta - \sin\theta = 2\sin\theta$$
$$\sqrt{3}\cos\theta = 3\sin\theta$$

Dividing each term by $3\cos\theta$ gives

$$\frac{\sqrt{3}\cos\theta}{3\cos\theta} = \frac{3\sin\theta}{3\cos\theta}$$

$$\frac{\sqrt{3}}{3} = \tan\theta$$

$$\tan\theta = \frac{\sqrt{3}}{3} \times \frac{\sqrt{3}}{\sqrt{3}} = \frac{1}{\sqrt{3}}, \text{ as required}$$

Note:
It is permissible to divide by $\cos\theta$ here, since $\cos\theta = 0$ is clearly not a solution of the equation.

Recall:
$\dfrac{\sin\theta}{\cos\theta} = \tan\theta$

Tip:
Manipulate the surd to get the required format.

Step 4: Use the result in **a** to solve the simple trig equation.

b $\cos(\theta + 30°) = \sin\theta$

$$\Rightarrow \quad \tan\theta = \frac{1}{\sqrt{3}}$$

$$\theta = 30°, 210°$$

Tip:
The values in range are PV and PV + 180°. Remember that the tan function repeats every 180°.

Example 2.7 Prove the identity

$$\cos A \cos(A - B) + \sin A \sin(A - B) \equiv \cos B$$

Step 1: Expand the LHS using the addition formulae.

$$\begin{aligned} \text{LHS} &= \cos A(\cos A \cos B + \sin A \sin B) \\ &\quad + \sin A(\sin A \cos B - \cos A \sin B) \\[4pt] &= \cos^2 A \cos B + \cos A \sin A \sin B \\ &\quad + \sin^2 A \cos B - \sin A \cos A \sin B \end{aligned}$$

Step 2: Simplify and use appropriate identities to arrive at the RHS.

$$= \cos^2 A \cos B + \sin^2 A \cos B$$

$$= \cos B(\cos^2 A + \sin^2 A)$$

$$= \cos B$$

$$= \text{RHS}$$

So $\cos A \cos(A - B) + \sin A \sin(A - B) \equiv \cos B$.

Tip:
You will not gain any marks if all you do is quote the formulae for $\cos(A - B)$ and $\sin(A - B)$ given in the formulae booklet. You must then go on to use them in the question.

Tip:
$\cos^2 A + \sin^2 A \equiv 1$.

Tip:
Keep going, even when the expression looks very complicated. Often terms cancel or you can use other identities to simplify further.

Double angle formulae

Putting $B = A$ in the identities for the sum of two angles gives the **double angle formulae** which are true for all values of A.

$$\sin 2A \equiv 2\sin A \cos A$$

$$\cos 2A \equiv \cos^2 A - \sin^2 A$$

$$\tan 2A \equiv \frac{2\tan A}{1 - \tan^2 A}$$

Note:
You should learn these. However, if you forget them, put $B = A$ in the formulae for $\sin(A + B)$, $\cos(A + B)$ and $\tan(A + B)$, given in the formulae booklet.

Often, alternative formats of the formula for $\cos 2A$ are used. Using $\sin^2 A + \cos^2 A \equiv 1$

$$\cos 2A \equiv \cos^2 A - \sin^2 A$$

$$\equiv \cos^2 A - (1 - \cos^2 A)$$

$$\equiv 2\cos^2 A - 1$$

Also,

$$\cos 2A \equiv \cos^2 A - \sin^2 A$$

$$\equiv (1 - \sin^2 A) - \sin^2 A$$

$$\equiv 1 - 2\sin^2 A$$

Note:
Do not expect always to see $2A$ and A as the angles. Look out especially for *half angles*, where, for example

$$\sin A \equiv 2\sin\frac{A}{2}\cos\frac{A}{2}$$

$$\cos A \equiv 2\cos^2\frac{A}{2} - 1$$

Summarising:

$$\cos 2A \equiv \cos^2 A - \sin^2 A \equiv 2\cos^2 A - 1 \equiv 1 - 2\sin^2 A$$

Note:
You must be confident about using any of these formats.

Example 2.8 Solve the equation $\cos 2x + 3 \sin x = 2$ for $0° \leqslant x < 360°$.

Step 1: Use an appropriate double angle formula.

$$\cos 2x + 3 \sin x = 2$$
$$\Rightarrow \quad 1 - 2 \sin^2 x + 3 \sin x = 2$$
$$2 \sin^2 x - 3 \sin x + 1 = 0$$

Step 2: Solve the equation in $\sin x$.

$$(2 \sin x - 1)(\sin x - 1) = 0$$
$$\Rightarrow \quad 2 \sin x - 1 = 0 \qquad \text{or } \sin x - 1 = 0$$
$$\sin x = 0.5 \qquad\qquad \sin x = 1$$
$$x = 30°, 150° \qquad\qquad x = 90°$$

So $x = 30°, 90°, 150°$

> **Tip:**
> Form an equation in $\sin x$ by using the version of $\cos 2x$ that involves $\sin x$.

> **Tip:**
> PV = 30°; the other solution in range is 180° − PV.

> **Tip:**
> PV = 90°. This is the only value in range. Be careful not to include other incorrect values within the range which will lead to loss of marks.

Example 2.9 Given that $\tan x = \frac{3}{4}$, using an appropriate double angle formula find the *exact* value of $\cot 2x$.

Step 1: Expand using an appropriate double angle formula.

First find the value of $\tan 2x$.

$$\tan 2x \equiv \frac{2 \tan x}{1 - \tan^2 x}$$

Step 2: Substitute the known value.

$$= \frac{2 \times \frac{3}{4}}{1 - \left(\frac{3}{4}\right)^2} = \frac{24}{7}$$

Step 3: Use the reciprocal function.

$$\cot 2x = \frac{1}{\tan 2x} = 1 \div \frac{24}{7} = \frac{7}{24}$$

> **Tip:**
> You will gain no marks for calculating arctan $\left(\frac{3}{4}\right)$ and using the angle obtained to calculate $\cot 2x$.

> **Tip:**
> If you use the fraction key on the calculator, you may need to put $\frac{3}{4}$ in a bracket before squaring.

Example 2.10 For values of x in the interval $0 \leqslant x \leqslant 2\pi$, solve the following equations, giving your answers exactly in terms of π.

a $\sin x = \sin 2x$ **b** $\sin\left(\frac{1}{2}x\right) = \sin x$

a

Step 1: Use an appropriate double angle formula.

$$\sin x = \sin 2x$$
$$\Rightarrow \quad \sin x = 2 \sin x \cos x$$
$$2 \sin x \cos x - \sin x = 0$$

Step 2: Factorise and solve.

$$\sin x(2 \cos x - 1) = 0$$
$$\Rightarrow \quad \sin x = 0 \qquad \text{or } 2 \cos x - 1 = 0$$
$$x = 0, \pi, 2\pi \qquad\qquad \cos x = 0.5$$
$$x = \tfrac{1}{3}\pi, \tfrac{5}{3}\pi$$

So $x = 0, \tfrac{1}{3}\pi, \pi, \tfrac{5}{3}\pi, 2\pi$

> **Tip:**
> Do not divide through by $\sin x$ as this will result in the loss of some solutions. Take it out as a factor.

Step 3: Notice the link with part **a** and apply a suitable substitution.

b Letting $\frac{1}{2}x = \theta$, the equation

$$\Rightarrow \quad \sin\left(\tfrac{1}{2}x\right) = \sin x, \qquad 0 \leqslant x \leqslant 2\pi$$

becomes $\quad \sin \theta = \sin 2\theta \qquad 0 \leqslant \theta \leqslant \pi$

Step 4: Solve using your answers from **a**.

From **a**, $\qquad \theta = 0, \tfrac{1}{3}\pi, \pi$
$$\Rightarrow \quad \tfrac{1}{2}x = 0, \tfrac{1}{3}\pi, \pi$$
$$x = 0, \tfrac{2}{3}\pi, 2\pi$$

> **Tip:**
> Be on the look out for links between parts of questions.

> **Tip:**
> Double your answers from part **a**, but only include values in the appropriate range.

Example 2.11 **a** By expanding $\sin(2A + A)$ prove that $\sin 3A \equiv 3 \sin A - 4 \sin^3 A$.

   b Hence, or otherwise, solve $3 \sin x - 4 \sin^3 x = 1$ for values of x such that $0° < x < 180°$.

Step 1: Use the addition formula suggested.

a $\sin 3A \equiv \sin(2A + A)$

$\equiv \sin 2A \cos A + \cos 2A \sin A$

Step 2: Use appropriate formulae to express all terms in terms of $\sin A$.

$\equiv 2 \sin A \cos A \cos A + (1 - 2 \sin^2 A)\sin A$

$\equiv 2 \sin A \cos^2 A + \sin A - 2 \sin^3 A$

$\equiv 2 \sin A(1 - \sin^2 A) + \sin A - 2 \sin^3 A$

$\equiv 2 \sin A - 2 \sin^3 A + \sin A - 2 \sin^3 A$

$\equiv 3 \sin A - 4 \sin^3 A$

> **Recall:**
> Formulae for $\sin(A + B)$, $\sin 2A$ and $\cos 2A$, $\cos^2 A$.

Step 3: Use the result in **a** to form a simple equation and solve in the appropriate range.

b $3 \sin x - 4 \sin^3 x = 1$

$\Rightarrow \sin 3x = 1$ Range for $3x$:

$\Rightarrow \quad 3x = 90°, 450°$ $0° < x < 180°$

$\quad\quad x = 30°, 150°$ $0° < 3x < 540°$

> **Tip:**
> Make sure that you have included all the solutions in range.

Example 2.12 Prove that $\dfrac{\sin 2A}{1 - \cos 2A} \equiv \cot A$.

Step 1: Use an appropriate double angle formula in the numerator.

$\text{LHS} = \dfrac{\sin 2A}{1 - \cos 2A}$

$= \dfrac{2 \sin A \cos A}{1 - \cos 2A}$

Step 2: Use an appropriate double angle formula in the denominator.

$= \dfrac{2 \sin A \cos A}{1 - (1 - 2 \sin^2 A)}$

$= \dfrac{2 \sin A \cos A}{2 \sin^2 A}$

$= \dfrac{\cos A}{\sin A}$

$= \cot A$

$= \text{RHS}$

So $\dfrac{\sin 2A}{1 - \cos 2A} \equiv \cot A$.

> **Tip:**
> In the numerator, use $\sin 2A = 2 \sin A \cos A$.

> **Tip:**
> In the denominator, use $\cos 2A = 1 - 2 \sin^2 A$.

> **Tip:**
> Do your working in stages, one thing at a time. Do not try to do too many things at once.

SKILLS CHECK **2B: Addition and double angle formulae**

1 Given that angles A and B are acute and that $\sin A = \frac{4}{5}$ and $\cos B = \frac{12}{13}$, find the exact values of:

   a $\cos A$ **b** $\sin B$ **c** $\cos(A - B)$ **d** $\sec(A - B)$

2 **a** Show that $\sin(x + 45°) = \frac{1}{\sqrt{2}} (\sin x + \cos x)$.

   b Hence solve the equation $\sin(x + 45°) = \sqrt{2} \cos x$ for $0° \leqslant x \leqslant 360°$.

3 a Simplify

    i $\sin(A + B) + \sin(A - B)$ **ii** $\cos(A + B) + \cos(A - B)$

  b Hence prove the identity $\dfrac{\sin(A + B) + \sin(A - B)}{\cos(A + B) + \cos(A - B)} \equiv \tan A$

4 Find the values of θ, where $0° \leqslant \theta \leqslant 360°$, such that

  a $\cos 2\theta = 1 + \sin \theta$ **b** $\sin 2\theta = \cot \theta$

5 Find the values of x, where $0 \leqslant x \leqslant 2\pi$, such that

  a $\cos 2x = \cos x$ **b** $\cos x = \cos \frac{1}{2}x$

6 a Prove the identity $\tan A + \cot A \equiv 2 \operatorname{cosec} 2A$.

  b Hence, for $0 < x < 2\pi$, solve the equation $\tan x + \cot x = 8$, giving your answer in radians correct to two decimal places.

7 a Express $\cos 2A$ in terms of

    i $\cos A$ **ii** $\sin A$

  b Prove that   **i** $\dfrac{2\cos \theta - \sec \theta}{\operatorname{cosec} \theta - 2\sin \theta} \equiv \tan \theta.$ **ii** $8\sin^2\left(\dfrac{\theta}{2}\right)\cos^2\left(\dfrac{\theta}{2}\right) \equiv 1 - \cos 2\theta$

8 a Expand $\sin(X - Y)$.

  b By letting $X = 4A$ and $Y = 2A$, or otherwise, prove the identity

$$\frac{\sin 4A \cos 2A - \cos 4A \sin 2A}{\sin A} \equiv 2 \cos A$$

9 Given that $\tan(A + B) = 1$ and $\tan A = \frac{1}{3}$, find the value of $\tan B$.

10 a Prove by a counter-example that the statement '$\cot(A + B) \equiv \cot A + \cot B$ for all A and B' is false.

  b Prove that $\cot(A + B) \equiv \dfrac{\cot A \cot B - 1}{\cot A + \cot B}.$

SKILLS CHECK **2B EXTRA** is on the CD

2.4 Expressions for $a \cos \theta + b \sin \theta$

Knowledge and use of expressions for $a\cos \theta + b\sin \theta$ in the equivalent forms of $r\cos(\theta \pm \alpha)$ or $r\sin(\theta \pm \alpha)$.

The expression $a\cos \theta + b\sin \theta$ can be written in the form $r\cos(\theta \pm \alpha)$ or $r\sin(\theta \pm \alpha)$.

Note:
This form is useful for solving equations, drawing graphs and finding maximum and minimum values of functions.

Example 2.13 It is given that $f(\theta) = 2\cos \theta + 3\sin \theta$.

  a Express $f(\theta)$ in the form $r\cos(\theta - \alpha)$, where $r > 0$ and $0° < \alpha < 90°$.

  b Hence solve the equation $2\cos \theta + 3\sin \theta = 3$, where $0° < \theta < 360°$. If an answer is not exact, give it correct to one decimal place.

Step 1: Use an appropriate identity to expand the trig expression.

a

$$r\cos(\theta - \alpha) \equiv 2\cos\theta + 3\sin\theta$$
$$\Rightarrow \quad r(\cos\theta\cos\alpha + \sin\theta\sin\alpha) \equiv 2\cos\theta + 3\sin\theta$$
$$r\cos\theta\cos\alpha + r\sin\theta\sin\alpha \equiv 2\cos\theta + 3\sin\theta$$

Recall:
$\cos(A - B)$
$\equiv \cos A\cos B + \sin A\sin B$

Step 2: Equate coefficients to form two equations that enable you to work out r and α.

Equating coefficients of $\cos\theta$:
$$r\cos\alpha = 2 \qquad \qquad \text{①}$$
Equating coefficients of $\sin\theta$:
$$r\sin\alpha = 3 \qquad \qquad \text{②}$$

Step 3: Divide the equations to find α.

Equation ② ÷ equation ① gives
$$\frac{r\sin\alpha}{r\cos\alpha} = \frac{3}{2}$$
$$\Rightarrow \quad \tan\alpha = \tfrac{3}{2}$$
$$\alpha = 56.30\ldots°$$

Recall:
$\dfrac{\sin\alpha}{\cos\alpha} \equiv \tan\alpha$

Tip:
$\alpha = \arctan\left(\dfrac{b}{a}\right) = \arctan\left(\dfrac{3}{2}\right)$

Step 4: Square and add the equations to find r.

Squaring ① and ② and adding gives
$$r^2\cos^2\alpha + r^2\sin^2\alpha = 2^2 + 3^2$$
$$r^2(\cos^2\alpha + \sin^2\alpha) = 13$$
$$r^2 = 13$$
$$r = \sqrt{13}$$

Recall:
$\cos^2\alpha + \sin^2\alpha \equiv 1$

Tip:
$r = \sqrt{a^2 + b^2} = \sqrt{2^2 + 3^2}$

Step 5: Write $f(\theta)$ in the required format.

So $f(\theta) = 2\cos\theta + 3\sin\theta \equiv \sqrt{13}\cos(\theta - 56.30\ldots°)$

Step 6: Use the format found in **a** to form a simple trig equation and solve.

b
$$2\cos\theta + 3\sin\theta = 3$$
$$\Rightarrow \quad \sqrt{13}\cos(\theta - 56.30\ldots°) = 3$$
$$\cos(\theta - 56.30\ldots°) = \frac{3}{\sqrt{13}} = 0.8320\ldots$$

Let $x = \theta - 56.30\ldots°$.

The equation becomes $\cos x = 0.8320\ldots$

Now $0° < \theta < 360°$

so $-56.30° < x < 303.69°$

PV of $x = 33.69\ldots°$

Tip:
Consider the range required for x. In this case you will need to consider negative values.

Note:
The next positive value is $360° - 33.69\ldots° = 326.30\ldots°$ which is out of range.

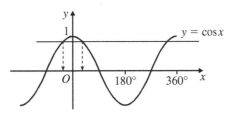

The other value in range is $-33.69\ldots°$.

So $\quad \theta - 56.30\ldots° = -33.69\ldots°, 33.69\ldots°$
$$\theta = 22.6° \text{ (1 d.p.)}, 90°$$

Example 2.14 It is given that $f(x) = 3\sin x - 3\cos x$.

a Find exact values of r and α such that $f(x) \equiv r\sin(x - \alpha)$, where $r > 0$ and $0 < \alpha < \tfrac{1}{2}\pi$.

b The diagram shows a sketch of $y = f(x)$ for $0 \le x \le 2\pi$. It crosses the y-axis at A and the x-axis at B and C. It has a maximum point at D and a minimum point at E.

  i Describe the transformations that map the curve $y = \sin x$ onto the curve $y = f(x)$.

  ii Find the exact coordinates of points A, B, C, D and E.

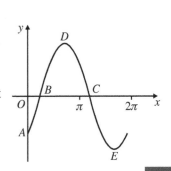

a

$$r \sin(x - \alpha) \equiv 3 \sin x - 3 \cos x$$
$$\Rightarrow \quad r(\sin x \cos \alpha - \cos x \sin \alpha) \equiv 3 \sin x - 3 \cos x$$
$$r \sin x \cos \alpha - r \cos x \sin \alpha \equiv 3 \sin x - 3 \cos x$$

Equating coefficients of $\sin x$:
$$r \cos \alpha = 3 \qquad\qquad ①$$
Equating coefficients of $\cos x$:
$$r \sin \alpha = 3 \qquad\qquad ②$$

Equation ② ÷ equation ① gives
$$\frac{r \sin \alpha}{r \cos \alpha} = \frac{3}{3}$$
$$\Rightarrow \qquad \tan \alpha = 1$$
$$\alpha = \tfrac{1}{4}\pi$$

Squaring ① and ② and adding gives
$$r^2 \cos^2 \alpha + r^2 \sin^2 \alpha = 3^2 + 3^2$$
$$r^2 (\cos^2 \alpha + \sin^2 \alpha) = 18$$
$$r^2 = 18$$
$$r = \sqrt{18} = \sqrt{9 \times 2} = 3\sqrt{2}$$

So $\quad f(x) = 3\sqrt{2} \sin(x - \tfrac{1}{4}\pi)$, with $r = 3\sqrt{2}$ and $\alpha = \tfrac{1}{4}\pi$.

b i To transform the curve $y = \sin x$ to $y = 3\sqrt{2} \sin(x - \tfrac{1}{4}\pi)$, translate by $\tfrac{1}{4}\pi$ units in the positive x-direction and stretch by $3\sqrt{2}$ in the y-direction.

ii $y = 3\sqrt{2} \sin(x - \tfrac{1}{4}\pi)$

When $x = 0$, $y = 3\sqrt{2} \sin(-\tfrac{1}{4}\pi) = 3\sqrt{2} \times (-\tfrac{1}{\sqrt{2}}) = -3$
So the coordinates of A are $(0, -3)$.

When $y = 0$, $3\sqrt{2} \sin(x - \tfrac{1}{4}\pi) = 0$
$$\Rightarrow \qquad x - \tfrac{1}{4}\pi = 0 \qquad \text{or } x - \tfrac{1}{4}\pi = \pi$$
$$\Rightarrow \qquad x = \tfrac{1}{4}\pi \qquad\qquad x = \tfrac{5}{4}\pi$$
So B is the point $(\tfrac{1}{4}\pi, 0)$ and C is the point $(\tfrac{5}{4}\pi, 0)$.

The maximum value of $\sin x$ is 1 and it occurs when $x = \tfrac{1}{2}\pi$.
Hence the maximum value of $3\sqrt{2} \sin(x - \tfrac{1}{4}\pi)$ is $3\sqrt{2} \times 1 = 3\sqrt{2}$, and it occurs when $x - \tfrac{1}{4}\pi = \tfrac{1}{2}\pi \Rightarrow x = \tfrac{3}{4}\pi$.
So D is the point $(\tfrac{3}{4}\pi, \ 3\sqrt{2})$.

The minimum value of $\sin x$ is -1 and it occurs when $x = \tfrac{3}{2}\pi$.
Hence the minimum value of $3\sqrt{2} \sin(x - \tfrac{1}{4}\pi)$ is
$3\sqrt{2} \times (-1) = -3\sqrt{2}$, and it occurs when $x - \tfrac{1}{4}\pi = \tfrac{3}{2}\pi \Rightarrow x = \tfrac{7}{4}\pi$.
So E is the point $(\tfrac{7}{4}\pi, -3\sqrt{2})$.

Which alternative format is preferable?

If the alternative format of $a \cos \theta + b \sin \theta$ is not specified in the question, it is advisable to ensure that α is acute, that is $0° < \alpha < 90°$ or $0 < \alpha < \tfrac{1}{2}\pi$.

For example, you could use the following alternatives:
For $\quad 3 \cos \theta - 4 \sin \theta$ use $r \cos(\theta + \alpha)$
For $\quad 4 \sin \theta - 3 \cos \theta$ use $r \sin(\theta - \alpha)$

Recall:
$\sin(A - B)$
$\equiv \sin A \cos B - \cos A \sin B$

Recall:
$\dfrac{\sin \alpha}{\cos \alpha} = \tan \alpha$

Tip:
Exact values are required, so write the angle in terms of π.

Recall:
$\cos^2 \alpha + \sin^2 \alpha = 1$

Tip:
Leave your answer in surd form. It is good practice to simplify it.

Recall:
Transformations (Section 1.6).

Recall:
$\sin 0 = \sin \pi = 0$

Recall:
$\sin \tfrac{1}{2}\pi = 1$

Recall:
$\sin \tfrac{3}{2}\pi = -1$

Tip:
In the cos format, the sign in the middle is opposite to the sign in the expression. In the sin format, the sign in the middle is the same as the sign in the expression.

However for $3 \cos \theta + 4 \sin \theta$ you could use either format:
$$3 \cos \theta + 4 \sin \theta \equiv r \cos(\theta - \alpha)$$
or $\qquad 4 \sin \theta + 3 \cos \theta \equiv r \sin(\theta + \alpha)$

SKILLS CHECK 2C: Alternative expressions for $a \cos \theta + b \sin \theta$

1 a Express $2 \cos x + \sin x$ in the form $r \cos(x - \alpha)$, where $r > 0$ and $0° < \alpha < 90°$.

   b Hence solve the equation $2 \cos x + \sin x = 1$, for values of x in the interval $0° < x < 360°$.

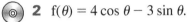 **2** $f(\theta) = 4 \cos \theta - 3 \sin \theta$.

   a Express $f(\theta)$ in the form $r \cos(\theta + \alpha)$, where $r > 0$ and $0° < \alpha < 90°$, giving the value of α in degrees, correct to 1 decimal place.

   b Hence

      i write down the maximum value of $f(\theta)$,

      ii find the largest negative value of θ at which $f(\theta)$ is maximum.

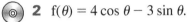 **3 a** Express $\sin x + \cos x$ in the form $r \sin(x + \alpha)$, where $r > 0$ and $0 < \alpha < \frac{1}{2}\pi$, giving the exact value of α.

   b Hence show that one of the solutions of the equation $\sin x + \cos x = \frac{1}{\sqrt{2}}$ is $x = \frac{7}{12}\pi$ and find the exact value of the other solution in the interval $-\pi < x < \pi$.

4 a Express $3 \sin x - 4 \cos x$ in the form $R \sin(x - \alpha)$, where $R > 0$ and $0° < \alpha < 90°$.

   b Describe how the graph of $y = 3 \sin x - 4 \cos x$ can be obtained from the graph of $y = \sin x$ by applying appropriate transformations.

5 The expression $k \cos x + 15 \sin x$ can be written in the form $17 \cos(x - \alpha)$, where $0° < \alpha < 90°$ and $k > 0$. Find the values of α and k.

SKILLS CHECK 2C EXTRA is on the CD

Examination practice 2: Trigonometry

1 Solve
$$6 \cos x = 1 + \sec x$$
for $0° < x < 360°$, giving answers correct to one decimal place where necessary.

2 Solve, showing clear working and giving your answers in radians to two decimal places,
$$6 \sec^2 2x + 5 \tan 2x = 12, \qquad 0 \leqslant x \leqslant \pi.$$
                                                              [Edexcel June 2000]

3 Solve, giving your answers in terms of π,
$$\cos 2x + 3 \sin x = 2, \qquad 0 \leqslant x < 2\pi.$$
                                                            [London June 1999]

4 Find the values of x in the interval $0 < x < 270$ which satisfy the equation
$$\frac{\tan 2x° + \tan 40°}{1 - \tan 2x° \tan 40°} = 1.$$
                                                            [London Jan 1998]

5 a Express $6\sin x \cos x + 4\sin^2 x - 2$ in the form $a\sin 2x + b\cos 2x$, where a and b are constants to be found.

b Hence solve $6\sin x \cos x + 4\sin^2 x - 2 = 0$ for $0° < x < 360°$.

6 Find, in terms of π, all solutions in the interval $0 \leqslant x < 2\pi$ of

a $\sin 2x = \sqrt{2}\cos x$,

b $2\sin\left(2x + \dfrac{\pi}{3}\right) = \cos\left(2x - \dfrac{\pi}{6}\right)$.

[Edexcel Jan 2001]

7 Given that $\sin(x + \alpha) = \sqrt{2}\cos(x - \alpha)$, where $\cos x \cos \alpha \neq 0$,

a prove that $\tan x = \dfrac{\sqrt{2} - \tan \alpha}{1 - \sqrt{2}\tan \alpha}$.

b Hence, or otherwise, find the solutions in the interval $0 < x < 2\pi$ of

$$\sin\left(x + \frac{\pi}{6}\right) = \sqrt{2}\cos\left(x - \frac{\pi}{6}\right),$$

giving your answers in radians to 3 decimal places.

[London Jan 2000]

8 Prove that $\dfrac{1 + \tan^2 \theta}{1 - \tan^2 \theta} \equiv \sec 2\theta$.

9 a Express $7\cos \theta + 24\sin \theta$ in the form $r\cos (\theta - \alpha)$, where $r > 0$ and $0 < \alpha < \dfrac{\pi}{2}$.

b Hence, or otherwise, solve the equation

$$7\cos \theta + 24\sin \theta = 12.5,$$

for $0 < \theta < 2\pi$, giving your answers to 1 decimal place.

10 a Express $\sin x + \sqrt{3}\cos x$ in the form $R\sin(x + \alpha)$, where $R > 0$ and $0 < \alpha < 90°$.

b Show that the equation $\sec x + \sqrt{3}\operatorname{cosec} x = 4$ can be written in the form

$$\sin x + \sqrt{3}\cos x = 2\sin 2x.$$

c Deduce from parts **a** and **b** that $\sec x + \sqrt{3}\operatorname{cosec} x = 4$ can be written in the form

$$\sin 2x - \sin(x + 60°) = 0.$$

d Hence, using the identity $\sin X - \sin Y \equiv 2\cos \dfrac{X + Y}{2} \sin \dfrac{X - Y}{2}$, or otherwise, find the values

of x in the interval $0 \leqslant x \leqslant 180°$, for which $\sec x + \sqrt{3}\operatorname{cosec} x = 4$.

[Edexcel Jan 2003]

11 The diagram shows an isosceles triangle ABC with $AB = AC = 4$ cm and $\angle BAC = 2\theta$.

The mid-points of AB and AC are D and E respectively. Rectangle $DEFG$ is drawn, with F and G on BC. The perimeter of rectangle $DEFG$ is P cm.

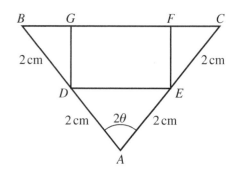

a Show that $DE = 4\sin \theta$.

b Show that $P = 8\sin \theta + 4\cos \theta$.

c Express P in the form $R\sin(\theta + \alpha)$,

where $R > 0$ and $0 < \alpha < \dfrac{\pi}{2}$.

Given that $P = 8.5$,

d find to 3 significant figures, the possible values of θ.

[Edexcel Jan 2005]

3 Exponentials and logarithms

3.1 The function e^x and its graph

The function e^x and its graph.

In *Core 2* you met graphs of the form $y = a^x$, known as exponential curves.

Recall that

- when $x = 0$, $y = a^0 = 1$ so the graph passes through $(0, 1)$

- for $a > 1$, as $x \to \infty$, $y \to \infty$ and as $x \to -\infty$, $y \to 0$, so the x-axis is an asymptote.

Recall:

Exponent means index or power (C1 Section 1.1).

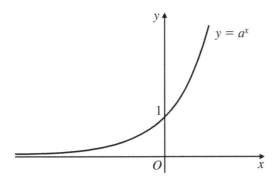

Recall:

An asymptote is a line that, as x tends to a particular value, the curve approaches but never meets (C1 Section 1.11).

Remember that the graph of a^x becomes steeper than the polynomial graphs you have studied.

The function $f(x) = e^x$, where e is the irrational number $2.718...$, is called **the** exponential function.

At the point $(0, 1)$, where $y = e^x$ cuts the y-axis, the curve has gradient 1.

Note:

e^x is sometimes written $\exp x$.

Note:

In fact, for all points on the curve $y = e^x$, $\dfrac{dy}{dx} = e^x$

(Section 4.1).

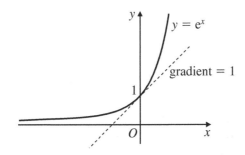

$e^x > 0$ for all values of x and, using the laws of indices, $e^0 = 1$.

Notice that as $x \to \infty$, $e^x \to \infty$ rapidly.

Recall:

$a^0 = 1$ (C1 Section 1.1).

Example 3.1 It is given that $f(x) = 2 + e^{-x}$.

a Sketch the graph of $y = f(x)$.

b State the equations of any asymptotes to the curve.

c State the domain and range of the function.

Tip:

The word 'state' means you don't need to calculate these answers or show any working: you should just be able to write them down.

Step 1: Decide the general shape of the curve, using your knowledge of transformations.

a Let $y = 2 + e^{-x}$.

The graph of $y = e^{-x}$ is a reflection in the y-axis of $y = e^x$.

The graph of $y = 2 + e^{-x}$ is a translation of $y = e^{-x}$ by 2 units in the y-direction.

Step 2: Set $x = 0$ to find the y-intercept.

When $x = 0$, $y = 2 + e^0 = 2 + 1 = 3$, so the curve cuts the y-axis at $(0, 3)$.

Step 3: Sketch the curve, marking the intercept.

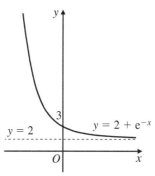

Recall:
$y = f(-x)$ is a reflection in the y-axis of $y = f(x)$, $y = f(x) + a$ translates the graph of $y = f(x)$ by a units in the y-direction (C1 Section 1.13).

Note:
e^{-x} is always greater than 0, so y has to be greater than $2 + 0$, i.e. $y > 2$.

Tip:
As $x \to \infty$, $f(x) \to 2$, so don't let your graph turn upwards at the end.

Step 4: Write down the equation of the asymptote.

b The x-axis is an asymptote to $y = e^{-x}$, so $y = 2$ is an asymptote to the translated curve.

Step 5: Using the graph, write down the domain and range.

c The domain is $x \in \mathbb{R}$ and the range is $y \in \mathbb{R}$, $y > 2$.

Recall:
The domain is the set of values that x can take. The range is the set of values that y can take. Both can be determined from the graph (Section 1.2).

3.2 The function ln *x* and its graph

The function ln x and its graph; ln x as the inverse of e^x.

In *Core 2* you used the fact that $x = a^y \Leftrightarrow \log_a x = y$ to show that logarithmic functions are the inverses of exponential functions.

It follows that $x = e^y \Leftrightarrow \log_e x = y$.

The logarithm to the base e of x is written **ln x** and is the **natural** log of x.

Since $\ln x$ is the inverse of e^x, the graph of $y = \ln x$ is a reflection of the graph of $y = e^x$ in the line $y = x$.

The domain of $\ln x$ is the range of e^x, i.e. $x \in \mathbb{R}$, $x > 0$.
The range of $\ln x$ is the domain of e^x, i.e. $y \in \mathbb{R}$.

Note:
$x = e^y \Leftrightarrow \ln x = y$

Recall:
Inverse functions (Section 1.4).

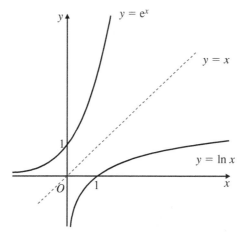

Using the laws of logs, $\ln 1 = 0$ and $\ln e = 1$.

Notice that as $x \to \infty$, $\ln x \to \infty$ slowly.

Recall:
$\log_a 1 = 0$, $\log_a a = 1$ (C2 Section 5.2).

Example 3.2 Describe the two transformations required to map the graph of $y = \ln x$ onto the graph of $y = \frac{1}{2}\ln(x - 2) + 1$.

Sketch the graph of $y = \frac{1}{2}\ln(x - 2) + 1$.

Step 1: Define the transformations.

The graph of $y = \frac{1}{2}\ln(x - 2) + 1$ is obtained from the graph of $y = \ln x$ by a stretch of scale factor $\frac{1}{2}$ parallel to the y-axis, followed by a translation by $\begin{pmatrix} 2 \\ 1 \end{pmatrix}$.

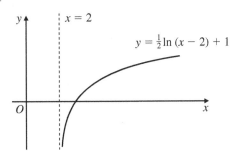
Recall:
$y = af(x)$ is a stretch, scale factor a in the y-direction.
$y = f(x - b) + c$ is a translation by $\begin{pmatrix} b \\ c \end{pmatrix}$ (Section 1.6).

Step 2: Consider any asymptotes.

Since $y = \ln x$ has an asymptote at $x = 0$, $y = \frac{1}{2}\ln(x - 2) + 1$ must have an asymptote at $x = 2$.

Step 3: Draw in the asymptote.

Step 4: Sketch the curve taking into account the stretch and translation.

Note:
The stretch must be carried out first. If the order of the transformations is reversed, the resulting graph is $y = \frac{1}{2}\ln(x - 2) + \frac{1}{2}$.

Example 3.3

Given that $f(x) = \ln(3 - x)$, $x \in \mathbb{R}$, $x < 3$, find the inverse function $f^{-1}(x)$, stating the domain and range of $f^{-1}(x)$.

Step 1: Let $y = f(x)$.

Let $y = \ln(3 - x)$

Step 2: Interchange x and y.

Now let $x = \ln(3 - y)$
$$e^x = 3 - y$$

Step 3: Make y the subject.

$$y = 3 - e^x$$
So $f^{-1}(x) = 3 - e^x$.

Step 4: State the domain and range of the inverse function.

The domain is $x \in \mathbb{R}$.

The range is $f^{-1}(x) \in \mathbb{R}$, $f^{-1}(x) < 3$.

Tip:
Be careful!
$\ln(3 - x) \neq \ln 3 - \ln x$

Recall:
$\ln y = x \Leftrightarrow y = e^x$

Note:
You can interchange x and y at the end if you prefer.

Recall:
The domain of f^{-1} is the range of f. The range of f^{-1} is the domain of f (Section 1.4).

Example 3.4

Sketch the graph of $y = \ln|x - 2|$, $x \neq 2$, stating the coordinates of the points where the graph meets the axes.

Step 1: Decide the general shape of the graph.

$y = \ln|x - 2|$
For $x > 0$, $y = \ln|x| = \ln x$.
For $x < 0$, $y = \ln|x| = \ln(-x)$, i.e. the curve to the right of the y-axis is reflected in the y-axis.
$y = \ln|x - 2|$ is a translation of $y = \ln|x|$ by 2 units in the x-direction.
$y = \ln|x|$ has the y-axis as an asymptote, so $y = \ln|x - 2|$ has $x = 2$ as an asymptote.

Recall:
For x −ve, $y = f(|x|)$ is a reflection of the curve to the right of the y-axis in the y-axis (Section 1.5).
$y = f(x + a)$ is a translation of $f(x)$ by $-a$ units in the x-direction (C1 Section 1.13).

Step 2: Set x and y equal to 0 to find any intersections with the axes.

When $x = 0$, $y = \ln|0 - 2| = \ln|-2| = \ln 2$, so the curve crosses the y-axis at $(0, \ln 2)$.

When $y = 0$, $\ln|x - 2| = 0$
$$|x - 2| = e^0 = 1$$

Step 3: Solve for both branches of the modulus.

Either $x - 2 = 1$ or $-(x - 2) = 1 \Rightarrow x - 2 = -1$
$\quad\quad x = 3 \quad\quad\quad\quad\quad\quad\quad\quad\quad x = 1$

So the curve crosses the x-axis at $(1, 0)$ and $(3, 0)$.

Tip:
Marks will be given for a symmetrical graph.

Step 4: Sketch the curve, marking the intercepts and the asymptote.

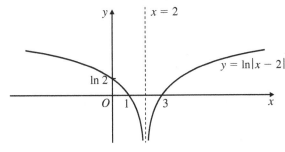

Tip:
It will help you to sketch the graph if you draw in the asymptote first, and there may be a mark for showing where it is.

Tip:
Make sure that your graph doesn't dip down at the end.

Solving equations

When solving equations involving e^x or $\ln x$, make use of the laws of indices and logarithms that you already know.

Recall:
Laws of indices (C1 Section 1.1), laws of logs (C2 Section 5.2).

You will also need to use the fact that $\ln x$ is the inverse of e^x.

- To solve an equation of the form $e^{ax+b} = p$, first take natural logs of both sides.

- To solve an equation of the form $\ln(ax + b) = q$, first rewrite it as $ax + b = e^q$.

Recall:
Natural logs are logs to the base e (Section 3.2).

Example 3.5 Find the exact solutions of the following equations:

a $e^{6x-1} = 3$ **b** $e^x = 6e^{-x} + 5$

c $\ln(2y + 1)^2 = 6$ **d** $\ln(y + 1) - \ln y = 2$

Tip:
The word 'exact' here means leave your answers in terms of $\ln a$ or e^a. Don't give your answer as a rounded decimal.

Step 1: Take natural logs of both sides.

a $e^{6x-1} = 3$

$\ln(e^{6x-1}) = \ln 3$

Recall:
$\log a^n = n \log a$ (C1 Section 1.1).

Step 2: Use an appropriate log law.

$(6x - 1)\ln e = \ln 3$

$6x - 1 = \ln 3$

Recall:
$\ln e = 1$

Step 3: Rearrrange the equation to find x.

$6x = \ln 3 + 1$

$x = \dfrac{\ln 3 + 1}{6}$

Note:
In questions where you are not asked for an exact solution, keep your answer in terms of $\ln a$ until the last line. That way you won't make any accuracy errors.

b $e^x = 6e^{-x} + 5$

Step 1: Substitute for e^x.

Let $y = e^x$

Note:
If $y = e^x$, then $e^{-x} = \dfrac{1}{e^x} = \dfrac{1}{y}$.

Then $y = 6 \times \dfrac{1}{y} + 5 = \dfrac{6}{y} + 5$

Step 2: Multiply through by y.

$y^2 = 6 + 5y$

Tip:
Don't forget to multiply **every** term by y.

Step 3: Rearrrange the equation and factorise.

$y^2 - 5y - 6 = 0$

$(y - 6)(y + 1) = 0$

Tip:
If the expression does not factorise use the quadratic formula.

Step 4: Solve for y.

$\Rightarrow \quad y = 6 \text{ or } y = -1$

Step 5: Substitute back e^x for y.

So $e^x = 6$ or $e^x = -1$.

Tip:
Don't forget to finish the question! Marks are commonly lost by candidates who forget to substitute back.

Step 6: Solve for x, using $\ln x$ as the inverse of e^x.

Since $e^x > 0$ for all values of x, $e^x = -1$ has no solution, but if $e^x = 6$, $x = \ln 6$.

c $\ln(2y + 1)^2 = 6$

Step 1: Use an appropriate log law.

$2 \ln(2y + 1) = 6$

$\ln(2y + 1) = \dfrac{6}{2} = 3$

Recall:
$\log a^n = n \log a$

Step 2: Change the log to exponential form.

$2y + 1 = e^3$

Tip:
$\ln y = x \Leftrightarrow y = e^x$

Step 3: Rearrange to find y.

$2y = e^3 - 1$

$y = \dfrac{e^3 - 1}{2}$

Note:
You don't need to simplify the log first; an alternative method would be $(2y + 1)^2 = e^6$, so $2y + 1 = \sqrt{e^6} = e^{\frac{6}{2}} = e^3$, as before.

d $\ln(y + 1) - \ln y = 2$

Step 1: Simplify using the log laws.

$$\ln\frac{y + 1}{y} = 2$$

Step 2: Change the log to exponential form.

$$\frac{y + 1}{y} = e^2$$

Step 3: Eliminate the denominator.

$$y + 1 = e^2y$$

$$e^2y - y = 1$$

Step 4: Make y the subject by rearranging and factorising.

$$y(e^2 - 1) = 1$$

$$y = \frac{1}{e^2 - 1}$$

> **Recall:**
>
> $\log x - \log y = \log\dfrac{x}{y}$ (C2 Section 5.2).

> **Tip:**
>
> Be careful! $\ln\dfrac{y + 1}{y} \neq \dfrac{\ln y + 1}{\ln y}$.

> **Tip:**
>
> From your calculator $\dfrac{1}{e^2 - 1} = 0.1565...$
>
> Substitute back into the original equation to make sure you haven't made a slip.

Simple exponential growth and decay

In real life, exponential growth (e.g. population growth, investment growth) is modelled around the exponential equation $y = Ae^{kt}$, where A and k are constants, $k > 0$.

> **Note:**
>
> Exponential growth and decay are studied in greater detail in C4.

Exponential decay (e.g. decay in radioactive isotopes, depreciation in value, temperature cooling) is modelled around the exponential equation $y = Ae^{-kt}$, where A and k are constants, $k > 0$.

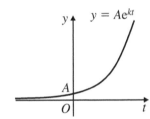
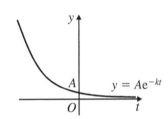

Example 3.6 The temperature, $T\,°C$, of a microwave meal t minutes after it has been heated is given by

$$T = 18 + 50\,e^{-\frac{t}{20}}, \qquad t \geqslant 0$$

a Find, in °C, the temperature of the meal the instant that it has been heated.

b Calculate, in °C to three significant figures, the temperature of the meal 5 minutes after it has been heated.

c Calculate, to the nearest minute, the time at which the temperature of the meal is 45 °C.

Step 1: Substitute $t = 0$ to find T.

a When $t = 0$,

$$T = 18 + 50\,e^{-\frac{0}{20}}$$

$$= 18 + 50\,e^0$$

$$= 18 + 50 \times 1$$

$$= 68$$

So the temperature of the meal the instant that it has been heated is 68 °C.

> **Tip:**
>
> t is the time after the meal has been heated, so $t = 0$ when the meal comes out of the oven.

> **Recall:**
>
> $e^0 = 1$

Step 2: Substitute the given value of t to find T.

b When $t = 5$,

$$T = 18 + 50\,\mathrm{e}^{-\frac{5}{20}}$$

$$= 18 + 50\,\mathrm{e}^{-\frac{1}{4}}$$

$$= 18 + 50 \times 0.7788...$$

$$= 56.94...$$

So the temperature of the meal five minutes after it has been heated is 56.9 °C (3 s.f.).

> **Tip:**
> Use the full value in your calculation, but don't forget to give your answer to the required degree of accuracy.

Step 3: Substitute the given value of T into the equation.

c When $T = 45$,

$$45 = 18 + 50\,\mathrm{e}^{-\frac{t}{20}}$$

$$27 = 50\,\mathrm{e}^{-\frac{t}{20}}$$

Step 4: Simplify the equation.

$$\frac{27}{50} = \mathrm{e}^{-\frac{t}{20}}$$

Step 5: Take natural logs of both sides and use an appropriate log law to simplify.

$$\ln\frac{27}{50} = \ln\mathrm{e}^{-\frac{t}{20}}$$

$$= -\frac{t}{20}\ln\mathrm{e}$$

$$= -\frac{t}{20}$$

> **Note:**
> You need to isolate the exponential term ($\mathrm{e}^{-\frac{t}{20}}$) before you take logs.

> **Recall:**
> $\ln\mathrm{e} = 1$

Step 6: Rearrange to find t and evaluate using your calculator.

$$-20\ln\frac{27}{50} = t$$

$$t = 12.3...$$

So the temperature of the meal is 45 °C approximately 12 minutes after it is heated.

> **Tip:**
> Use the exact values in your working; only use decimals at the end.

SKILLS CHECK   3A: Exponentials and logarithms

 1 Solve $\mathrm{e}^{-\frac{x}{3}} = \frac{1}{2}$, giving your answer in the form $a \ln b$ where a and b are integers.

 2 Solve $3\mathrm{e}^{2x} = 2(\mathrm{e}^x + 4)$.

3 Solve $\mathrm{e}^{2x-5} = 1$.

4 Find the exact solution of $\ln(4x + 3) = 0.5$.

5 Sketch the graph of $y = \ln|2x - 3|$, $x \in \mathbb{R}$, $x \neq \frac{3}{2}$.
Find the exact coordinates of any points of intersection of the graph with the axes.

6 For each of the functions
    a $\mathrm{f}(x) = 3\mathrm{e}^x$, $x \in \mathbb{R}$ **b** $\mathrm{f}(x) = \ln 2x$, $x \in \mathbb{R}$, $x > 0$
  i find the inverse function, $\mathrm{f}^{-1}(x)$,
  ii sketch the graphs of $y = \mathrm{f}(x)$ and $y = \mathrm{f}^{-1}(x)$ on the same axes, showing the coordinates of any points of intersection with the axes,
  iii state the range of $\mathrm{f}(x)$ and the domain and range of $\mathrm{f}^{-1}(x)$.

7 a Given that $\mathrm{f}(x) = \mathrm{e}^{2x+1}$, $x \in \mathbb{R}$, find the inverse function $\mathrm{f}^{-1}(x)$.
  b Sketch the graphs of $y = \mathrm{f}(x)$ and $y = \mathrm{f}^{-1}(x)$ on the same axes, stating the equations of any asymptotes.
  c State the range of $\mathrm{f}(x)$ and the domain and range of $\mathrm{f}^{-1}(x)$.

8 Given that $f(x) = \frac{1}{4}\ln(x + 1)$, $x \in \mathbb{R}$, $x > -1$,

    a find the inverse function, $f^{-1}(x)$,

    b state the domain and range of $f^{-1}(x)$,

    c find the exact value of x for which $f(x) = \frac{1}{2}$.

9 Sketch the graph of $y = e^{3x+2} + 1$.

10 The amount an initial investment of £1000 is worth after t years is given by A, where
$$A = 1000\,e^{0.09t}, \quad t \geqslant 0.$$

    a How much is the investment worth after 5 years?

    b After how many years will the investment have doubled in value?

    c Sketch the graph of A against t.

SKILLS CHECK **3A EXTRA** is on the CD

Examination practice 3: Exponentials and logarithms

1 Solve the following equations, giving the exact values of x and y.

    a $e^{-2x} = \frac{1}{16}$ **b** $\ln y - \ln(y - 1) = 1$

2 Find the exact solutions of the following equations.

    a $e^{3x+6} = 8$ **b** $\ln(2x - 4) = 3$

 3 a Sketch, on the same axes, the graphs with equations $y = 1 + e^{-x}$ and $y = 2|x + 4|$.

    b Write down the coordinates of any points where the graphs meet the axes.

    The graphs intersect at the point where $x = p$.

    c Show that $x = p$ is a root of the equation $e^{-x} - 2x - 7 = 0$.

4 A curve has equation $y = f(x)$ where the function f is given by
$$f : x \mapsto e^{x+1} - 2, x \in \mathbb{R}.$$

    a Sketch the curve and write down the exact coordinates of the points of intersection with the axes.

    b Find the inverse function of f.

    c State the domain and range of $f^{-1}(x)$.

5 a Sketch the graph of $y = k + \ln\dfrac{x}{2}$.

    b Find, in terms of k, the coordinates of the point of intersection with the x-axis.

    c Given that the curve crosses the x-axis at the point $\left(\dfrac{2}{e^2}, 0\right)$, show that $k = 2$.

 6 The function f is given by
$$f : x \mapsto \ln(3x - 6), x \in \mathbb{R}, x > 2.$$

    a Find $f^{-1}(x)$.

    b Write down the domain of f^{-1} and the range of f^{-1}.

    c Find, to three significant figures, the value of x for which $f(x) = 3$.

    The function g is given by
$$g : x \mapsto \ln|3x - 6|, x \in \mathbb{R}, x \neq 2.$$

    d Sketch the graph of $y = g(x)$.

    e Find the exact coordinates of all the points at which the graph of $y = g(x)$ meets the coordinate axes.

                                                   [Edexcel June 2004]

7 The function f is given by

$$f : x \mapsto \ln(4 - 2x), \, x \in \mathbb{R}, \, x < 2.$$

a Find an expression for $f^{-1}(x)$.

b Sketch the curve with equation $y = f^{-1}(x)$, showing the coordinates of the points where the curve meets the axes.

The function g is given by

$$g : x \mapsto 3^x, \, x \in \mathbb{R}.$$

c Find the value of x for which $g(x) = 1.5$, giving your answer to three decimal places.

d Evaluate $gf(1)$ to three decimal places. [Edexcel Specimen Paper]

8 A particular species of orchid is being studied. The population p at time t years after the study started is assumed to be

$$p = \frac{2800ae^{0.2t}}{1 + ae^{0.2t}}, \text{ where } a \text{ is a constant.}$$

Given that there were 300 orchids when the study started,

a show that $a = 0.12$,

b use the equation with $a = 0.12$ to predict the number of years before the population of orchids reaches 1850.

c Show that $p = \dfrac{336}{0.12 + e^{-0.2t}}$.

d Hence show that the population cannot exceed 2800. [Edexcel June 2005]

4 Differentiation

4.1 Differentiation of functions

Differentiation of e^x, ln x, sin x, cos x, tan x and their sums and differences. Differentiation using the chain rule.

In *Core 1* and *Core 2* you learnt how to differentiate expressions consisting of sums and differences of powers of x and some of the uses of differentiation.

Recall the following results:

$$y = ax^n \Rightarrow \frac{dy}{dx} = nax^{n-1}$$

and $\quad y = f(x) \pm g(x) \Rightarrow \frac{dy}{dx} = f'(x) \pm g'(x)$

This table shows the derivatives of some common functions needed in *Core 3*.

$f(x)$	$f'(x)$
e^x	e^x
$\ln x$	$\dfrac{1}{x}$
$\sin x$	$\cos x$
$\cos x$	$-\sin x$
$\tan x$	$\sec^2 x$

Recall:
Multiply by the power of x and decrease the power by 1 (C1, Section 4.2).

Recall:
You can differentiate term by term (C1 Section 4.2).

Note:
Remember that the trigonometric results are only true for x in **radians**.

Note:
The derivative of tan x can be deduced from the derivative of tan kx given in the formula booklet, or by using the quotient rule (Section 4.2).

Example 4.1 Differentiate with respect to x:

a $x^3 - 2\cos x$ **b** $\ln(4x^3)$

Step 1: Differentiate term by term.

a Let $y = x^3 - 2\cos x$

$$\frac{dy}{dx} = 3x^2 - 2(-\sin x)$$

Step 2: Tidy up the signs.

$$= 3x^2 + 2\sin x$$

Tip:
It can help to wait until the end to simplify the signs, rather than trying to do too many things at once.

Step 1: Simplify the expression using log laws.

b Let $y = \ln(4x^3)$

$$= \ln 4 + \ln x^3$$
$$= \ln 4 + 3\ln x$$

Step 2: Differentiate term by term.

$$\frac{dy}{dx} = 0 + 3 \times \frac{1}{x} = \frac{3}{x}$$

Note:
Where possible, simplify log terms before differentiating but note that $\ln(4x^3) \neq 3\ln(4x)$.
Use $\log(xy) = \log x + \log y$ and $\log x^a = a\log x$ (C2 Section 5.2).

Example 4.2 A curve C, with equation $y = x^2 + 3e^x - 1$, crosses the y-axis at the point P. Find an equation of the normal to C at P.

Step 1: Find $\dfrac{dy}{dx}$.

$$y = x^2 + 3e^x - 1$$
$$\frac{dy}{dx} = 2x + 3e^x$$

Step 2: Substitute the x-value to get the gradient of the tangent at P.

When $x = 0$, $\dfrac{dy}{dx} = 2(0) + 3e^0 = 3$

Recall:
$e^0 = 1$ (Section 3.1).

Tip:
Be careful when evaluating $3e^0$. It is 3, not 0 or 1.

Step 3: Find the gradient of the normal at P.

Step 4: Substitute x into the equation of the curve to get y.

Step 5: Use an appropriate straight line equation.

The gradient of the tangent at P is 3

$\Rightarrow$ gradient of normal at P is $-\frac{1}{3}$

When $x = 0$,

$y = 0^2 + 3e^0 - 1 = 3 - 1 = 2$

Equation of the normal at $P(0, 2)$:

$y - 2 = -\frac{1}{3}(x - 0)$

$y - 2 = -\frac{1}{3}x$

$3y - 6 = -x$

$\quad 3y = 6 - x$

> **Recall:**
> The tangent and normal are perpendicular, so the product of their gradients is -1 (C1 Section 2.2).

> **Recall:**
> Equation of line
> $y - y_1 = m(x - x_1)$
> (C1 Section 2.1).

> **Note:**
> Often you will be asked to give the equation in a specific format. If you are not, then leave it in a convenient form.

Chain rule

The chain rule is one of the most useful results in differentiation. It enables composite functions such as $(5x - 3)^7$, $4 \sin^2 x$, $\frac{1}{2}e^{3x + 5}$ and $\ln \dfrac{1}{\sqrt{2x + 9}}$ to be differentiated.

If y is a function of t, and t is a function of x, then, by the chain rule,

$$\frac{dy}{dx} = \frac{dy}{dt} \times \frac{dt}{dx}$$

An alternative version of the rule, using function notation, is

$$\frac{d}{dx}f(g(x)) = f'(g(x)) \times g'(x)$$

> **Recall:**
> A composite function is formed by combining two or more functions (Section 1.3).

> **Note:**
> Sometimes this is referred to as differentiating a function of a function.

Example 4.3 Find $\dfrac{dy}{dx}$ when

a $y = (5x - 3)^7$

b $y = 4 \sin^2 x$

c $y = \frac{1}{2}e^{3x + 5}$

d $y = \ln \dfrac{1}{\sqrt{2x + 9}}$

Step 1: Define t as a function of x; y is now a function of t.

Step 2: Differentiate t with respect to x, and y with respect to t.

Step 3: Rewrite t in terms of x.

Step 4: Apply the chain rule.

a $y = (5x - 3)^7$

Let $t = 5x - 3$, then $y = t^7$

So $\dfrac{dt}{dx} = 5$

and $\dfrac{dy}{dt} = 7t^6$

$\qquad = 7(5x - 3)^6$

$\dfrac{dy}{dx} = \dfrac{dy}{dt} \times \dfrac{dt}{dx}$

$\qquad = 7(5x - 3)^6 \times 5$

$\qquad = 35(5x - 3)^6$

Step 1: Define *t* as a function of *x*; *y* is now a function of *t*.

b $y = 4\sin^2 x = 4(\sin x)^2$

Let $t = \sin x$, then $y = 4t^2$

Step 2: Differentiate *t* with respect to *x*, and *y* with respect to *t*.

So $\dfrac{dt}{dx} = \cos x$

and $\dfrac{dy}{dt} = 2 \times 4t$

$= 8\sin x$

Step 3: Rewrite *t* in terms of *x*.

Step 4: Apply the chain rule.

$\dfrac{dy}{dx} = \dfrac{dy}{dt} \times \dfrac{dt}{dx}$

$= 8\sin x \times \cos x$

$= 8\sin x \cos x$

Step 1: Define *t* as a function of *x*; *y* is now a function of *t*.

c $y = \frac{1}{2}e^{3x + 5}$

Let $t = 3x + 5$, then $y = \frac{1}{2}e^t$

Step 2: Differentiate *t* with respect to *x*, and *y* with respect to *t*.

So $\dfrac{dt}{dx} = 3$

and $\dfrac{dy}{dt} = \frac{1}{2}e^t$

Step 3: Rewrite *t* in terms of *x*.

Step 4: Apply the chain rule.

$= \frac{1}{2}e^{3x + 5}$

$\dfrac{dy}{dx} = \dfrac{dy}{dt} \times \dfrac{dt}{dx}$

$= \frac{1}{2}e^{3x + 5} \times 3$

$= \frac{3}{2}e^{3x + 5}$

Step 1: Simplify the expression using log laws.

d $y = \ln \dfrac{1}{\sqrt{2x + 9}}$

$= \ln(2x + 9)^{-\frac{1}{2}}$

$= -\frac{1}{2}\ln(2x + 9)$

Step 2: Define *t* as a function of *x*; *y* is now a function of *t*.

Step 3: Differentiate *t* with respect to *x*, and *y* with respect to *t*.

Let $t = 2x + 9$, then $y = -\frac{1}{2}\ln t$

So $\dfrac{dt}{dx} = 2$

and $\dfrac{dy}{dt} = -\dfrac{1}{2} \times \dfrac{1}{t}$

Step 4: Rewrite *t* in terms of *x*.

$= -\dfrac{1}{2(2x + 9)}$

Step 5: Apply the chain rule.

$\dfrac{dy}{dx} = \dfrac{dy}{dt} \times \dfrac{dt}{dx}$

$= -\dfrac{1}{2(2x + 9)} \times 2$

$= -\dfrac{1}{(2x + 9)}$

Some standard derivatives, using the chain rule

The following results come from the chain rule and are useful to remember.

f(x)	f'(x)
e^{kx}	ke^{kx}
$\ln kx$	$\dfrac{1}{x}$
$\sin kx$	$k \cos kx$
$\cos kx$	$-k\sin kx$
$\tan kx$	$k\sec^2 kx$

Note:
Don't worry if you forget these; you can always work them out. The derivative of $\tan kx$ is given in the formula booklet.

Example 4.4 Given that $y = \sin^2 5x$, find $\dfrac{dy}{dx}$, writing your answer in the form $a \sin bx$, where a and b are integers.

Step 1: Define t as a function of x; y is now a function of t.

$y = \sin^2 5x = (\sin 5x)^2$

Let $t = \sin 5x$ then $y = t^2$

Step 2: Differentiate t with respect to x, and y with respect to t.

So $\dfrac{dt}{dx} = 5 \cos 5x$

Step 3: Rewrite t in terms of x.

and $\dfrac{dy}{dt} = 2t$

$= 2 \sin 5x$

Step 4: Apply the chain rule.

$\dfrac{dy}{dx} = \dfrac{dy}{dt} \times \dfrac{dt}{dx}$

Step 5: Use the double angle formula to write the answer in the required form.

$= 2 \sin 5x \times 5 \cos 5x$

$= 10 \sin 5x \cos 5x$

$= 5 \sin 10x$

Tip:
Don't worry about the required format for the answer until the end of the solution; what's needed will be much clearer then.

Tip:
If you had forgotten that $\dfrac{d}{dx}(\sin kx) = k\cos kx$ you would need to use the chain rule twice.

Recall:
$\sin 2A = 2 \sin A \cos A$ (Section 2.3).

Example 4.5 The curve C, with equation $y = 2e^{2x} - 4x$, has a stationary point at P. Find the coordinates of P and determine the nature of the point.

Step 1: Differentiate.

$y = 2e^{2x} - 4x$

$\dfrac{dy}{dx} = 2 \times 2e^{2x} - 4 = 4e^{2x} - 4$

Step 2: Put $\dfrac{dy}{dx} = 0$ and solve for x.

At P, $\quad \dfrac{dy}{dx} = 0$

so $\quad 4e^{2x} - 4 = 0$

$4(e^{2x} - 1) = 0$

$e^{2x} - 1 = 0$

$e^{2x} = 1$

Taking natural logs of both sides gives

$\ln e^{2x} = \ln 1$

$2x \ln e = 0$

$\Rightarrow \qquad 2x = 0$

$x = 0$

Tip:
If you had forgotten that $\dfrac{d}{dx}e^{kx} = ke^{kx}$ you could use the chain rule.

Recall:
Stationary points have zero gradient (C2 Section 6.1).

Recall:
$\ln x$ is the inverse of e^x; $\ln 1 = 0$; $\ln e = 1$ (Section 3.2).

Step 3: Substitute the x value into the equation of the curve to find y.

When $x = 0$, $y = 2e^{2(0)} - 4 \times 0 = 2 \times 1 - 0 = 2$, so P is the point $(0, 2)$.

Recall:
$e^0 = 1$ (Section 3.1).

Step 4: Find $\dfrac{d^2y}{dx^2}$.

$$\dfrac{d^2y}{dx^2} = 4 \times 2e^{2x} = 8e^{2x}$$

Step 5: Determine the sign of $\dfrac{d^2y}{dx^2}$ at the stationary point.

At P, $x = 0$, so $\dfrac{d^2y}{dx^2} = 8e^{2(0)} = 8 \times 1 = 8$.

Since $\dfrac{d^2y}{dx^2} > 0$, P is a minimum point.

Note:
If $\dfrac{d^2y}{dx^2} = 0$, then use the alternative method of testing the sign of $\dfrac{dy}{dx}$ either side of the stationary point.

Example 4.6 Find the stationary point of the curve $y = \tan^2 x - 2\tan x$ in the range $-\frac{1}{2}\pi < x < \frac{1}{2}\pi$.

Tip:
The way the question is worded implies there is only one stationary point in the range.

Step 1: Define t as a function of x; y is now a function of t.

$y = \tan^2 x - 2\tan x = (\tan x)^2 - 2\tan x$

Let $t = \tan x$, then $y = t^2 - 2t$

Note:
Remember the derivative of $\tan x$ is true for x in radians.

Step 2: Differentiate t with respect to x, and y with respect to t.

So $\dfrac{dt}{dx} = \sec^2 x$

and $\dfrac{dy}{dt} = 2t - 2$

Step 3: Rewrite t in terms of x.

$= 2\tan x - 2$

Step 4: Apply the chain rule.

$\dfrac{dy}{dx} = \dfrac{dy}{dt} \times \dfrac{dt}{dx}$

$= (2\tan x - 2) \times \sec^2 x$

$= 2\sec^2 x(\tan x - 1)$

Note:
Brackets are needed here.

Step 5: Put $\dfrac{dy}{dx} = 0$ and solve for x.

$\dfrac{dy}{dx} = 0$ when $2\sec^2 x(\tan x - 1) = 0$

So either $\sec^2 x = 0 \Rightarrow \sec x = 0$ (no solution)

or $\tan x - 1 = 0 \Rightarrow \tan x = 1$

$x = \frac{1}{4}\pi$

Recall:
$\sec x$ is never equal to 0 (Section 2.2). If you had forgotten this fact, you could have used $\sec^2 x = \tan^2 x + 1 = 0$ to get $\tan^2 x = -1$ which has no solutions.

Step 6: Substitute the x value into the equation of the curve to find y.

When $x = \frac{1}{4}\pi$,

$y = \tan^2(\frac{1}{4}\pi) - 2\tan(\frac{1}{4}\pi) = 1 - 2 = -1$

So the stationary point is at $\left(\frac{1}{4}\pi, -1\right)$.

Note:
If you use your calculator in radians mode, $x = 0.785$ (3 s.f.) would be an acceptable solution as the question didn't ask for an exact solution.

SKILLS CHECK **4A: Differentiation of functions**

1 Differentiate with respect to x:

   a $5e^x$ **b** $\ln x^3$ **c** $4\cos x - \sin x$ **d** $2x^3 - e^x$ **e** $\ln \dfrac{5}{x^2}$

2 Differentiate with respect to x:

   a $3\cos x^2$ **b** $3\cos^2 x$ **c** $3e^{x^2 - 6x}$ **d** $4\ln(5 - 2x)$ **e** $\tan^2 3x$

3 Simplify $y = \ln \dfrac{5x}{\sqrt[3]{2x + 7}}$, then differentiate to find $\dfrac{dy}{dx}$, writing your answer in its simplest form.

4 A curve has equation $y = 2\sin x + \cos 2x$. Find the coordinates of the stationary point in the range $0 < x < \frac{1}{2}\pi$ and determine its nature.

5 Find an equation for the tangent to the curve $y = \left(\frac{1}{2}x^2 - 1\right)^3$ at the point where $x = -2$.

6 The curve C has equation $y = x^2 - 8\ln x$, $x > 0$. Show that C has only one stationary point and find its coordinates.

7 A curve C has equation $y = 3e^x - 2\ln 5x$.

    a Find $\dfrac{dy}{dx}$.

    b Find an equation of the tangent to the curve at $x = 1$.

    c Find the exact y-coordinate of the point where this tangent crosses the y-axis. Give your answer in the form $a - \ln b$.

8 The curve C, with equation $y = \tan x - \sin x$, $-\frac{1}{2}\pi < x < \frac{1}{2}\pi$, has a stationary point at P. Find the coordinates of P.

SKILLS CHECK **4A EXTRA** is on the CD

4.2 Product and quotient rules

Differentiation using product and quotient rules.

Product rule

To differentiate a product of two functions (i.e. an expression formed by multiplying two functions together) use the **product rule**.

If $y = uv$, where u and v are functions of x, then,

$$\frac{dy}{dx} = u\frac{dv}{dx} + v\frac{du}{dx}$$

An alternative version of the rule, using function notation, is

$$\frac{d}{dx}(f(x)g(x)) = f'(x)g(x) + f(x)g'(x)$$

Note:
In words: first × derivative of second + second × derivative of first.

Note:
If you use the alternative version of the rule, your working will be in a different order, but the answer will be the same.

Example 4.7 Differentiate with respect to x, giving your answer in its simplest form:

    a $(3x^2 - 2)(1 - 4x^3)$ **b** $2x^4 \sin x$

Step 1: Define u and v in terms of x.

a $y = (3x^2 - 2)(1 - 4x^3)$

Let $u = 3x^2 - 2$, $v = 1 - 4x^3$

Step 2: Differentiate u and v with respect to x.

So $\dfrac{du}{dx} = 6x$ and $\dfrac{dv}{dx} = -12x^2$

Step 3: Apply the product rule.

$\dfrac{dy}{dx} = u\dfrac{dv}{dx} + v\dfrac{du}{dx}$

$= (3x^2 - 2) \times (-12x^2) + (1 - 4x^3) \times 6x$

Step 4: Simplify.

$= -12x^2(3x^2 - 2) + 6x(1 - 4x^3)$

$= 6x[-2x(3x^2 - 2) + (1 - 4x^3)]$

$= 6x(-6x^3 + 4x + 1 - 4x^3)$

$= 6x(1 + 4x - 10x^3)$

Note:
The derivative could be found by multiplying out the brackets and then differentiating without using the product rule.

Tip:
Even when you are not requested to do so, it is good practice to tidy up expressions that result from the product rule.

Step 1: Define *u* and *v* in terms of *x*.

b $y = 2x^4 \sin x$

Let $u = 2x^4$, $v = \sin x$

Step 2: Differentiate *u* and *v* with respect to *x*.

So $\dfrac{du}{dx} = 8x^3$ and $\dfrac{dv}{dx} = \cos x$

Step 3: Apply the product rule.

$\dfrac{dy}{dx} = u\dfrac{dv}{dx} + v\dfrac{du}{dx}$

Step 4: Simplify.

$= 2x^4 \times \cos x + \sin x \times 8x^3$

$= 2x^3(x \cos x + 4 \sin x)$

Example 4.8 The curve *C* has equation $y = e^x \cos x$, $0 \leqslant x < \frac{1}{2}\pi$. Show that the curve has a stationary point when $\tan x = 1$.

Step 1: Define *u* and *v* in terms of *x*.

$y = e^x \cos x$

Let $u = e^x$, $v = \cos x$

Step 2: Differentiate *u* and *v* with respect to *x*.

So $\dfrac{du}{dx} = e^x$ and $\dfrac{dv}{dx} = -\sin x$

Step 3: Apply the product rule.

$\dfrac{dy}{dx} = u\dfrac{dv}{dx} + v\dfrac{du}{dx}$

$= e^x \times (-\sin x) + \cos x \times e^x$

Step 4: Simplify.

$= e^x(\cos x - \sin x)$

Step 5: Set $\dfrac{dy}{dx} = 0$ and solve.

Stationary points occur when $\dfrac{dy}{dx} = 0$, i.e. when

$e^x(\cos x - \sin x) = 0$

$\Rightarrow \quad e^x = 0$ (no solution) or $\cos x - \sin x = 0$

$\Rightarrow \quad \cos x = \sin x$

$\Rightarrow \quad \tan x = 1$

Quotient rule

To differentiate a quotient of two functions (i.e. an expression formed by dividing one function by another) use the **quotient rule**.

If $y = \dfrac{u}{v}$, where *u* and *v* are functions of *x*, then

$$\frac{dy}{dx} = \frac{v\dfrac{du}{dx} - u\dfrac{dv}{dx}}{v^2}$$

An alternative version of the rule, using function notation, is

$$\frac{d}{dx}\left(\frac{f(x)}{g(x)}\right) = \frac{f'(x)g(x) - f(x)g'(x)}{(g(x))^2}$$

Example 4.9 Differentiate $\dfrac{e^{3x}}{x}$ with respect to *x*.

Step 1: Define *u* and *v* in terms of *x*.

$y = \dfrac{e^{3x}}{x}$

Let $u = e^{3x}$, $v = x$

Step 2: Differentiate *u* and *v* with respect to *x*.

So $\dfrac{du}{dx} = 3e^{3x}$ and $\dfrac{dv}{dx} = 1$

Step 3: Apply the quotient rule.
$$\frac{dy}{dx} = \frac{v\dfrac{du}{dx} - u\dfrac{dv}{dx}}{v^2}$$

Step 4: Simplify.
$$= \frac{x \times 3e^{3x} - e^{3x} \times 1}{x^2}$$

$$= \frac{e^{3x}(3x - 1)}{x^2}$$

> **Note:**
> You could differentiate this as a product of e^{3x} and x^{-1} to give
> $$\frac{dy}{dx} = 3e^{3x}x^{-1} + e^{3x}(-x^{-2}),$$
> which then simplifies to the same result.

Derivatives of the reciprocal trigonometric functions

The derivatives of the following trigonometric functions are given in the formula booklet.

$f(x)$	$f'(x)$
$\sec x$	$\sec x \tan x$
$\cot x$	$-\operatorname{cosec}^2 x$
$\operatorname{cosec} x$	$-\operatorname{cosec} x \cot x$

> **Tip:**
> Remember that the derivatives of trigonometric functions starting with 'co' have a negative sign.

Example 4.10 Use the derivatives of $\sin x$ and $\cos x$ to prove that the derivative of $\cot x$ is $-\operatorname{cosec}^2 x$.

> **Note:**
> The question asks for proof, so you need to show all the steps in your working.

Step 1: Write $\cot x$ in terms of $\sin x$ and $\cos x$.
$$y = \cot x = \frac{1}{\tan x} = \frac{\cos x}{\sin x}$$

Step 2: Define u and v in terms of x.
$$\text{Let} \quad u = \cos x, \qquad v = \sin x$$

Step 3: Differentiate u and v with respect to x.
$$\text{So} \quad \frac{du}{dx} = -\sin x \quad \text{and} \quad \frac{dv}{dx} = \cos x$$

Step 4: Apply the quotient rule.
$$\frac{dy}{dx} = \frac{v\dfrac{du}{dx} - u\dfrac{dv}{dx}}{v^2}$$

$$= \frac{\sin x \times (-\sin x) - \cos x \times \cos x}{\sin^2 x}$$

Step 5: Simplify.
$$= \frac{-\sin^2 x - \cos^2 x}{\sin^2 x}$$

$$= -\frac{(\sin^2 x + \cos^2 x)}{\sin^2 x}$$

$$= -\frac{1}{\sin^2 x}$$

$$= -\operatorname{cosec}^2 x$$

> **Recall:**
> $\sin^2 x + \cos^2 x \equiv 1$ (C2 Section 4.6).

> **Recall:**
> $\operatorname{cosec} x = \dfrac{1}{\sin x}$ (Section 2.2).

SKILLS CHECK 4B: Product and quotient rules

1 Using the product rule, differentiate with respect to x:

a $x^3(x + 4)^2$ **b** $\sin x \tan x$ **c** $e^{2x} x^4$ **d** $x \ln\sqrt{x + 3}$

2 Using the quotient rule, differentiate with respect to x:

a $\dfrac{3x^2}{x - 3}$ **b** $\dfrac{\cos x}{1 - \sin x}$ **c** $\dfrac{e^{\frac{x}{2}}}{2x^3}$ **d** $\dfrac{\ln(x + 1)}{x + 1}$

3 Differentiate with respect to x:

a $(x^2 + 3)^3(5x - 4)^5$ **b** $\cos^3 x \sin 3x$ **c** $\dfrac{2e^x - 1}{2e^x + 1}$ **d** $\dfrac{\sin 2x}{x^2}$

4 The curve C has the equation $y = \dfrac{\ln x}{x^2}$. Find the gradient of C at the point where $x = e$.

5 Find the stationary points of the curve $y = x^3 e^{-x}$.

 6 Given that $f(x) = \dfrac{x^2}{x - 3}$,

     a show that $f'(x) = \dfrac{x^2 - 6x}{(x - 3)^2}$,   **b** find $f''(x)$ in its simplest form.

 7 Find an equation for the normal to the curve $y = \dfrac{x^2}{1 + x^2}$ at the point where $x = 1$.

    8 The curve C has equation $y = 2x^{\frac{1}{2}} e^{-x}$. Find the x-coordinate of the stationary point of the curve.

SKILLS CHECK **4B EXTRA** is on the CD

4.3 Finding $\dfrac{dy}{dx}$ when $x = f(y)$

The use of $\dfrac{dy}{dx} = \dfrac{1}{\dfrac{dx}{dy}}$.

When a function is defined in terms of y instead of x, a variation of the chain rule enables the derivative to be found, using

$$\frac{dy}{dx} = \frac{1}{\dfrac{dx}{dy}}$$

So, if x is given as a function of y, differentiate x with respect to y to find $\dfrac{dx}{dy}$ and then work out the reciprocal to get $\dfrac{dy}{dx}$.

> **Note:**
> It is interesting to note that this rule does not apply to the second derivative, so
> $$\frac{d^2y}{dx^2} \neq \frac{1}{\dfrac{d^2x}{dy^2}}$$

Example 4.11 Given that $x = \sin 3y$, show that $\dfrac{dy}{dx} = \tfrac{1}{3} \sec 3y$.

Step 1: Differentiate x with respect to y.

$x = \sin 3y$

$\dfrac{dx}{dy} = 3 \cos 3y$

Step 2: Work out the reciprocal to find $\dfrac{dy}{dx}$.

$\dfrac{dy}{dx} = \dfrac{1}{\dfrac{dx}{dy}}$

$= \dfrac{1}{3 \cos 3y}$

Step 3: Write result in required format.

$= \dfrac{1}{3} \times \dfrac{1}{\cos 3y}$

$= \tfrac{1}{3} \sec 3y$

> **Recall:**
> $\dfrac{d}{dx}(\sin ky) = k \cos ky$
> (Section 4.1).

> **Recall:**
> $\dfrac{1}{\cos y} = \sec y$ (Section 2.2).

> **Note:**
> Usually questions will expect the answer in terms of y, since the original function was defined that way, but watch out for questions asking for the answer in terms of x.

Example 4.12 Consider the function $y = \ln x$. By first expressing x in terms of y, show that $\dfrac{d}{dx}(\ln x) = \dfrac{1}{x}$.

Step 1: Express x in terms of y.

$$y = \ln x$$
$$\Rightarrow \quad x = e^y$$

Step 2: Differentiate x with respect to y.

$$\frac{dx}{dy} = e^y$$

Step 3: Rewrite in terms of x.

$$= x$$

Step 4: Work out the reciprocal to find $\dfrac{dy}{dx}$.

$$\frac{dy}{dx} = \frac{1}{\dfrac{dx}{dy}}$$

$$= \frac{1}{x}$$

i.e. $\dfrac{d}{dx}(\ln x) = \dfrac{1}{x}$

Recall:
e^x is the inverse of $\ln x$ (Section 3.2).

SKILLS CHECK **4C: Finding $\dfrac{dy}{dx}$ when $x = f(y)$**

1 Given that $x = \dfrac{y^2 - 5}{2}$, find $\dfrac{dy}{dx}$ in terms of y.

2 Find $\dfrac{dy}{dx}$ in terms of y, given that

 a $x = \sin y$ **b** $x = \cos y$.

3 Given that $x = \dfrac{y^2 + y}{y - 1}$, find $\dfrac{dy}{dx}$ in terms of y.

4 Given that $x = y^4$, find $\dfrac{dy}{dx}$ in terms of x.

5 Given the curve $x = \dfrac{y^2 + 2}{6}$,

 a find $\dfrac{dy}{dx}$,

 b find an equation for the normal to the curve at the point where $y = 1$.

6 The curve C has equation $x = \dfrac{y}{e^y}$. Find the gradient of the tangent to C at the point where $y = \ln 3$.

7 Given the curve $x = \dfrac{y^2}{4}$, for $x > 0$, $y > 0$,

 a find equations for the tangents to the curve at $x = 1$ and at $x = 4$,

 b find the coordinates of the point of intersection of these two tangents.

8 Given the curve C, with equation $x = (y^2 + 25)^{\frac{1}{2}}$, find an equation for the tangent to C at the point $(13, 12)$.

SKILLS CHECK **4C EXTRA is on the CD**

1 The curve C has equation $y = 4x^{\frac{3}{2}} - \ln(5x)$, where $x > 0$. The tangent at the point on C where $x = 1$ meets the x-axis at the point A.

Prove that the x-coordinate of A is $\frac{1}{5}\ln(5e)$.   [Edexcel Mock Examination]

2 The curve C has equation $y = 2e^x + 3x^2 + 2$. The point A with coordinates $(0, 4)$ lies on C. Find the equation of the tangent to C at A.   [Edexcel June 2001]

3 The curve C with equation $y = p + qe^x$, where p and q are constants, passes through the point $(0, 2)$. At the point $P(\ln 2, p + 2q)$ on C, the gradient is 5.

   a Find the value of p and the value of q.

   The normal to C at P crosses the x-axis at L and the y-axis at M.

   b Show that the area of $\triangle OLM$, where O is the origin, is approximately 53.8.   [Edexcel Nov 2002]

4 Given that $y = \log_a x$, $x > 0$, where a is a positive constant,

   a **i** express x in terms of a and y,

   ii deduce that $\ln x = y \ln a$.

   b Show that $\dfrac{dy}{dx} = \dfrac{1}{x \ln a}$.

   The curve C has equation $y = \log_{10} x$, $x > 0$. The point A on C has x-coordinate 10. Using the result in part **b**,

   c find an equation for the tangent to C at A.

   The tangent to C at A crosses the x-axis at the point B.

   d Find the exact x-coordinate of B.   [Edexcel Jan 2004]

5 It is given that $f(x) = x + \dfrac{2}{x - 2} - \dfrac{10}{x^2 + x - 6}$, $x \in \mathbb{R}$, $x < -3$.

   a Show that $f(x) = \dfrac{x^2 + 3x + 2}{x + 3}$.

   b Find $f'(x)$.

   c Hence solve the equation $f'(x) = \frac{7}{8}$.

6 **a** Given that $a^x \equiv e^{kx}$, where a and k are constants, $a > 0$ and $x \in \mathbb{R}$, prove that $k = \ln a$.

   b Hence, using the derivative of e^{kx}, prove that when $y = 2^x$, $\dfrac{dy}{dx} = 2^x \ln 2$.

   c Hence deduce that the gradient of the curve with equation $y = 2^x$ at the point $(2, 4)$ is $\ln 16$.   [Edexcel June 2001]

7 Differentiate with respect to x

   i $x^3 e^{3x}$,   **ii** $\dfrac{2x}{\cos x}$,   **iii** $\tan^2 x$.

   Given that $x = \cos y^2$,

   iv find $\dfrac{dy}{dx}$ in terms of y.   [Edexcel Mock Examination]

8 As a substance cools, its temperature, $T\,^{\circ}$C, is related to the time (t minutes) for which it has been cooling. The relationship is given by the equation

$$T = 20 + 60\,e^{-0.1t},\ t \geqslant 0.$$

a Find the value of T when the substance started to cool.

b Explain why the temperature of the substance is always above $20\,^{\circ}$C.

c Sketch the graph of T against t.

d Find the value, to 2 significant figures, of t at the instant $T = 60$.

e Find $\dfrac{dT}{dt}$.

f Hence find the value of T at which the temperature is decreasing at a rate of $1.8\,^{\circ}$C per minute.

[Edexcel Specimen Paper]

 9 The curve $y = \frac{1}{2}\ln(3x + 5)$ crosses the axes at $P(0, p)$ and $Q(q, 0)$.

a Find the exact values of p and q.

b Find an equation for the normal to the curve at Q, giving your answer in the form $ay + bx + c = 0$, where a, b and c are integers.

10 Given that $x = y^2 \ln y$, $y > 0$,

a find $\dfrac{dx}{dy}$,

b use your answer to part **a** to find, in terms of e, the value of $\dfrac{dy}{dx}$ at $y = $ e. [Edexcel Jan 2001]

5 Numerical methods

5.1 Location of roots

Location of the roots of f(x) = 0 by considering changes of sign in an interval of x in which f(x) is continuous.

If the function f(x) is continuous in the interval $a \leqslant x \leqslant b$, and f(x) changes sign in this interval, then f(x) = 0 has a root in the interval $a \leqslant x \leqslant b$.

Note:
$a \leqslant x \leqslant b$ can be written as [a, b].

Example 5.1 It is given that $f(x) = \sin x - 2x + 1$.
Show that f(x) = 0 has a root in the interval $0.8 \leqslant x \leqslant 0.9$.

Step 1: Substitute the boundary values of the interval into f(x) and consider the signs.

$f(0.8) = \sin 0.8 - 2 \times 0.8 + 1 = 0.117 \ldots > 0$
$f(0.9) = \sin 0.9 - 2 \times 0.9 + 1 = -0.016 \ldots < 0$

Step 2: Look for a sign change.

Change of sign $\Rightarrow$ root of f(x) = 0 lies in the interval $0.8 \leqslant x \leqslant 0.9$.

Note:
Use radians for trigonometric functions.

Tip:
Be careful. The interval is
$0.8 \leqslant x \leqslant 0.9$, **not**
$0.8 \leqslant f(x) \leqslant 0.9$.

Example 5.2 **a** Sketch, on the same axes, the graphs of $y = e^x$ and $y = 5 - x$.

Given that $f(x) = e^x + x - 5$,

b show that the equation f(x) = 0 has only one root,

c show that this root lies in the interval (1, 2).

Note:
(a, b) means $a < x < b$.

Step 1: Sketch the graphs. **a**

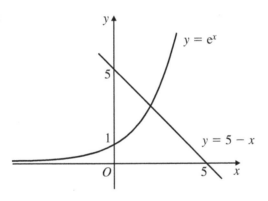

Recall:
Graph of e^x (Section 3.1).

Step 2: Consider the intersection point(s) of the two graphs.

b When the graphs intersect:
$$e^x = 5 - x$$
$$\Rightarrow \quad e^x + x - 5 = 0$$

From the sketch, it can be seen that the graphs of $y = 5 - x$ and $y = e^x$ have only one point of intersection
$\Rightarrow$ there is only one root of $e^x + x - 5 = 0$.
Hence the equation f(x) = 0 has only one root.

Tip:
Show your full method in case you make a slip.

Step 3: Substitute the boundary values of the interval into f(x) and consider the signs.

c $f(x) = e^x + x - 5$
$f(1) = e^1 + 1 - 5 = -1.28 \ldots < 0$
$f(2) = e^2 + 2 - 5 = 4.38 \ldots > 0$

Change of sign $\Rightarrow$ root of f(x) = 0 in (1, 2).

Tip:
You must state 'change of sign' and make a comment about there being a root.

1 Show that the equation $f(x) = 0$ has a root in the given interval.

a $f(x) = \sqrt[3]{x} + x - 7$ $\qquad$ $5 < x < 6$

b $f(x) = \cos 2x + x$ $\qquad$ $[-1, 0]$

c $f(x) = \ln(x - 4) + \sqrt{x}$ $\qquad$ $4.1 \leqslant x \leqslant 4.2$

d $f(x) = \tan x - e^x$ $\qquad$ $-4 < x < -3$

e $f(x) = \dfrac{1}{x} + 1 - x^3$ $\qquad$ $(1.2, 1.3)$

 2 a On the same axes, sketch the graphs of $y = \ln x$ and $y = x^2 - 4$.

b Hence write down the number of roots of the equation $\ln x - x^2 + 4 = 0$.

c Show that the equation $\ln x - x^2 + 4 = 0$ has a root in the interval $2 \leqslant x \leqslant 3$.

 3 a Using the same axes, sketch the graphs of $y = e^x - 1$ and $y = 2x + 1$.

b Hence show that the equation $e^x - 2x - 2 = 0$ has one negative root and one positive root.

The positive root of the equation $e^x - 2x - 2 = 0$ lies in the interval $n < x < n + 1$, where $n \in \mathbb{Z}$.

c Find the value of n.

SKILLS CHECK **5A EXTRA is on the CD**

5.2 Approximate solutions of equations

Approximate solutions of equations using simple iterative methods, including recurrence relations of the form $x_{n+1} = f(x_n)$.

An approximate solution of an equation can usually be found by rearranging the equation into the form $x_{n+1} = f(x_n)$ and then using this iterative formula, with an appropriate starting value, to find subsequent approximations.

This method fails if the sequence of values does not converge.

Note:
A converging sequence approaches a limit.

The accuracy of an approximate solution can be tested by using the change of sign method. For example, if a solution of the equation $f(x) = 0$ is 4.327 to three decimal places, then $f(4.3265)$ and $f(4.3275)$ will have different signs.

Example 5.3 a Show that the equation $x^3 + 3x - 7 = 0$ can be rearranged in the form $x = \sqrt[3]{7 - 3x}$.

b Use the iterative formula $x_{n+1} = \sqrt[3]{7 - 3x_n}$, with $x_0 = 1.4$, to find x_1, x_2, x_3, x_4 and x_5, giving your answers to three decimal places.

Note:
An equation of the form $x_{n+1} = f(x_n)$ is also called a recurrence relation.

Step 1: Rearrange the equation to the required format.

a $x^3 + 3x - 7 = 0$

$$x^3 = 7 - 3x$$
$$x = \sqrt[3]{7 - 3x}$$

Step 2: Apply the iterative formula. **b** $x_0 = 1.4$

$x_1 = \sqrt[3]{7 - 3(1.4)} = 1.4094\ldots = 1.409$ (3 d.p.)

$x_2 = \sqrt[3]{7 - 3(1.4094\ldots)} = 1.4046\ldots = 1.405$ (3 d.p.)

$x_3 = \sqrt[3]{7 - 3(1.4046\ldots)} = 1.4070\ldots = 1.407$ (3 d.p.)

$x_4 = \sqrt[3]{7 - 3(1.4070\ldots)} = 1.4058\ldots = 1.406$ (3 d.p.)

$x_5 = \sqrt[3]{7 - 3(1.4058\ldots)} = 1.4064\ldots = 1.406$ (3 d.p.)

Tip:
Make sure you use the given iterative formula.

Tip:
Use the full calculator display each time.

Tip:
If the values are not converging, check your working.

Calculator note:

If your calculator has the facility of reproducing the last answer, try keying in the following:

| 1.4 | = | $\sqrt[3]{}$ | (| 7 | − | 3 | Ans |) |

The successive terms of the iteration can then be obtained by keying in

| = | = | = | etc.

Note:
On graphical calculators, press
| EXE | or | ENTER |

Example 5.4 It is given that $f(x) = \ln x + x + 4$, $x > 0$.

a Show that $f(x) = 0$ has a root in the interval $(0.01, 0.02)$.

b Show that $f(x) = 0$ can be rearranged in the form $x = e^{-(x + 4)}$.

c Using the iterative formula $x_{n+1} = e^{-(x_n + 4)}$ with $x_1 = 0.01$, find the values of x_2, x_3 and x_4, giving the value of x_4 to four decimal places.

d By considering the change of sign of $f(x)$ in a suitable interval, show that your value for x_4 gives an accurate estimate, correct to four decimal places, of the root of $f(x) = 0$.

Tip:
Even if you can't answer **b**, you could still do parts **c** and **d** as they are not dependent on previous answers.

Step 1: Substitute the boundary values of the interval into $f(x)$ and consider the signs. **a** $f(x) = \ln x + x + 4$

$f(0.01) = \ln 0.01 + 0.01 + 4 = -0.595\ldots < 0$

$f(0.02) = \ln 0.02 + 0.02 + 4 = 0.107\ldots > 0$

Change of sign $\Rightarrow$ root of $f(x) = 0$ in $(0.01, 0.02)$.

Tip:
Write down the value you get for your calculation to make it clear whether it is positive or negative.

Step 2: Rearrange the equation to the required format. **b** $\qquad f(x) = 0$

$\ln x + x + 4 = 0$

$\ln x = -x - 4 = -(x + 4)$

$\Rightarrow x = e^{-(x + 4)}$

Recall:
$\ln x = y \Leftrightarrow x = e^y$ (Section 3.2).

Step 3: Apply the iterative formula. **c** $x_1 = 0.01$

$x_2 = e^{-(0.01 + 4)} = 0.01813\ldots$

$x_3 = e^{-(0.01813\ldots + 4)} = 0.01798\ldots$

$x_4 = e^{-(0.01798\ldots + 4)} = 0.01798\ldots = 0.0180$ (4 d.p.)

Tip:
Since you are asked for the root to 4 d.p. you must work to at least 5 d.p. Here the full calculator display has been used each time.

Calculator note:

Try keying in the following:

| 0.01 | = | e^x | (−) | (| Ans | + | 4 |) |

| = | = | = | etc.

d
$$f(x) = \ln x + x + 4$$

$$f(0.01795) = \ln(0.01795) + 0.01795 + 4 = -0.00221... < 0$$

$$f(0.01805) = \ln(0.01805) + 0.01805 + 4 = 0.00344... > 0$$

Change of sign $\Rightarrow$ root of $f(x) = 0$ is in $(0.01795, 0.01805)$

$\Rightarrow$ root is 0.0180 (correct to four decimal places)

Tip:
If the root is 0.0180, correct to four decimal places, then it must lie in the interval (0.01795, 0.01805).

Divergence

Sometimes an iterative process does not lead to a root of the equation, even if you use a starting value close to that root. The sequence x_0, x_1, x_2, ... may diverge.

Example 5.5 **a** Show that the equation $\frac{1}{2}x^3 - 2x + 1 = 0$ has a root in the interval $(1, 2)$.

b Use the iterative formula $x_{n+1} = \dfrac{x^3 + 2}{4}$ with $x_0 = 1.75$ to find x_1, x_2, x_3, x_4 and x_5, giving your answers to three decimal places.

c Comment on your sequence of values.

Step 1: Substitute the boundary values of the interval into f(x) and consider the signs.

a Let $f(x) = \frac{1}{2}x^3 - 2x + 1$

$$f(1) = \tfrac{1}{2}(1^3) - 2 \times 1 + 1 = -\tfrac{1}{2} < 0$$

$$f(2) = \tfrac{1}{2}(2^3) - 2 \times 2 + 1 = 1 > 0$$

Change of sign $\Rightarrow$ root of $f(x) = 0$ in $(1, 2)$.

Step 2: Apply the iterative formula.

b $x_{n+1} = \dfrac{x^3 + 2}{4}$

$$x_0 = 1.75$$

$$x_1 = \frac{1.75^3 + 2}{4} = 1.8398... = 1.840 \text{ (3 d.p.)}$$

$$x_2 = \frac{1.8398...^3 + 2}{4} = 2.0569... = 2.057 \text{ (3 d.p.)}$$

$$x_3 = \frac{2.0569...^3 + 2}{4} = 2.6758... = 2.676 \text{ (3 d.p.)}$$

$$x_4 = \frac{2.6758...^3 + 2}{4} = 5.2899... = 5.290 \text{ (3 d.p.)}$$

$$x_5 = \frac{5.2899...^3 + 2}{4} = 37.5070... = 37.507 \text{ (3 d.p.)}$$

Note:
The root is actually $x = 1.675$. Even though the iterative process started close to that root, the sequence is getting further and further away.

Calculator note:
Try keying in the following:

| 1.75 | = | (| Ans | x^3 | + | 2 |) |

| ÷ | 4 | = | = | = | etc.

Tip:
If you are using a calculator to generate your iterations be careful with brackets and make sure you write down all your intermediate results.

Step 3: Comment on the sequence.

c The sequence diverges.

1 In each of the following:

   i Show that the equation can be rearranged into the given iterative formula.

   ii Use the value of x_0 to find the values of x_1, x_2, x_3 and x_4, giving the value of x_4 to the stated accuracy.

a $x^2 - \dfrac{2}{x} - 1 = 0$ $x_{n+1} = \sqrt{\dfrac{2}{x_n} + 1}$ $x_0 = 1.5$ 3 decimal places

b $e^x - 2x - 5 = 0$ $x_{n+1} = \ln(2x_n + 5)$ $x_0 = 2.25$ 4 decimal places

c $\cos x - 9x - 4 = 0$ $x_{n+1} = \frac{1}{9}(\cos x_n - 4)$ $x_0 = -0.3$ 5 decimal places

d $4x^3 - x - 6 = 0$ $x_{n+1} = \sqrt[3]{\dfrac{x_n + 6}{4}}$ $x_0 = 1.2$ 4 significant figures

e $\ln x + 2 - \sqrt{x} = 0$ $x_{n+1} = e^{\sqrt{x_n} - 2}$ $x_0 = 0.2$ 3 decimal places

2 In each of the following:

   i Use the given iteration and the value of x_1 to obtain x_2, x_3 and x_4 to the stated accuracy.

   ii By considering the change of sign of $f(x)$ in an appropriate interval, show that your value of x_4 is the root of $f(x) = 0$ correct to the stated accuracy.

a $f(x) = \tan x - 3x$ $x_{n+1} = \arctan(3x_n)$ $x_1 = 1.32$ 3 decimal places

b $f(x) = x^3 - 3x + 4$ $x_{n+1} = \sqrt[3]{3x_n - 4}$ $x_1 = -2.2$ 3 decimal places

c $f(x) = x + \ln 5x$ $x_{n+1} = 0.2e^{-x_n}$ $x_1 = 0.16$ 3 decimal places

d $f(x) = x^3 - 11x^2 - 2$ $x_{n+1} = \dfrac{2}{x_n^2} + 11$ $x_1 = 10$ 5 significant figures

e $f(x) = e^{0.3x} - x - 2$ $x_{n+1} = \frac{10}{3}\ln(x_n + 2)$ $x_1 = 7.51$ 4 significant figures

 3 $f(x) = \cos x - x^2 + 3$.

   a Show that $f(x) = 0$ has a root in the interval $[1, 2]$.

   b Using the iterative formula $x_{n+1} = \sqrt{\cos x_n + 3}$ and $x_0 = 1.7$, write down the values of x_1, x_2 and x_3, giving your answer to x_3 to three decimal places.

4 **a** On the same axes, sketch the curves with equations $y = 2^x$ and $y = x^3 - 7$.

   b Use your sketch to show that the equation $2^x - x^3 + 7 = 0$ has exactly one solution.

   c Use the iterative formula $x_{n+1} = \sqrt[3]{2^{x_n} + 7}$ and $x_0 = 2.3$ to find x_1, x_2 and x_3.

   d Hence write down an approximate solution of the equation $2^x - x^3 + 7 = 0$.

 5 Given that $f(x) = x^2 - 4x - 8$,

   a show that $f(x) = 0$ has a root in the interval $[5, 6]$.

   b Use the iterative formula $x_{n+1} = \frac{1}{4}x_n^2 - 2$ and $x_0 = 6$ to find x_1, x_2 and x_3.

   c Comment on your sequence of results.

1 $f(x) = x^3 + \dfrac{2}{x} - 4, x \neq 0$

   a Show that $f(x) = 0$ has a solution in the interval $0.5 \leqslant x \leqslant 0.6$.

The solution is to be estimated using the iterative formula $x_{n+1} = \dfrac{2}{4 - x_n^3}, x_1 = 0.52$.

   b Calculate the values of x_2, x_3 and x_4, giving your answers to four significant figures.

   c Using a change of sign method over a suitable interval, show that the solution of $f(x) = 0$ is 0.5180 correct to four significant figures.

2 $f(x) = 0.1x - \ln(x + 2), x > -2$

   a Rearrange the equation $f(x) = 0$ into the form $x = e^{ax} + b$, stating the values of the constants a and b.

   b Use the iterative formula $x_{n+1} = e^{ax_n} + b$, with $x_1 = -1.1$ and your values of a and b to find x_2, x_3 and x_4.

   c Hence write down an approximation to the negative root of the equation $f(x) = 0$.

3 $f(x) = \cot\left(x - \dfrac{\pi}{3}\right) - \dfrac{1}{x}, x \neq 0$.

   a Given that $f(\alpha) = 0$, show that $2 < \alpha < 3$.

   b Use the iterative formula $x_{n+1} = \arctan x_n + \dfrac{\pi}{3}$, with $x_0 = 2.19$, to find x_1, x_2 and x_3.

   c Hence write down a suitable approximation for the value of α.

   d Use the change of sign method to justify the accuracy of your answer.

4 The diagram shows part of the curve with equation $y = f(x)$, where $f(x) = 3x^2 + e^{-x}$. The curve has a minimum at the point A.

   a Find $f'(x)$.

   b Show that the x-coordinate of A lies in the interval $0.1 < x < 0.2$.

A more accurate estimate of the x-coordinate of A is made using the iterative formula

     $x_{n+1} = \frac{1}{6}e^{-x_n}$ with $x_0 = 0.1$.

   c Write down the values of x_1, x_2, x_3 and x_4, giving the value of x_4 to three decimal places.

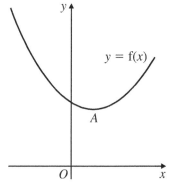

5 $f(x) \equiv 5x - 4\sin x - 2$, where x is in radians.

   a Evaluate, to 2 significant figures, $f(1.1)$ and $f(1.15)$.

   b State why the equation $f(x) = 0$ has a root in the interval $[1.1, 1.15]$.

An iteration formula of the form $x_{n+1} = p\sin x_n + q$ is applied to find an approximation to the root of the equation $f(x) = 0$ in the interval $[1.11, 1.15]$.

   c Stating the values of p and q, use this iteration formula with $x_0 = 1.1$ to find x_4 to 3 decimal places. Show the intermediate results of your working. [London June 1998]

6 $f(x) \equiv e^{0.8x} - \dfrac{1}{3 - 2x}, \; x \neq \dfrac{3}{2}.$

    a Show that the equation $f(x) = 0$ can be written as

$$x = 1.5 - 0.5e^{-0.8x}.$$

    b Use the iteration formula

$$x_{n+1} = 1.5 - 0.5e^{-0.8x_n},$$

    with $x_0 = 1.3$, to obtain x_1, x_2 and x_3. Give the value of x_3, an approximation to a root of $f(x) = 0$, to 3 decimal places.

    c Show that the equation $f(x) = 0$ can be written in the form $x = p \ln(3 - 2x)$, stating the value of p.

    d Use the iteration formula

$$x_{n+1} = p \ln(3 - 2x_n),$$

    with $x_0 = -2.6$ and the value of p found in part **c**, to obtain x_1, x_2 and x_3. Give the value of x_3, an approximation to the second root of $f(x) = 0$, to 3 decimal places.   [London Jan 2000]

7 $f(x) = x^3 - 2 - \dfrac{1}{x}, \; x \neq 0.$

    a Show that the equation $f(x) = 0$ has a root between 1 and 2.
      An approximation for this root is found using the iteration formula

$$x_{n+1} = \left(2 + \frac{1}{x_n}\right)^{\frac{1}{3}}, \text{ with } x_0 = 1.5.$$

    b By calculating the values of x_1, x_2, x_3 and x_4, find an approximation to this root, giving your answer to 3 decimal places.

    c By considering the change of sign of $f(x)$ in a suitable interval, verify that your answer to part **b** is correct to 3 decimal places.   [Edexcel Nov 2004]

8 **a** On the same axes, sketch the graphs of $y = \ln x$ and $y = 4 - x^2$.

    Given $f(x) = \ln x + x^2 - 4, \; x > 0,$

    b use your graphs to show that the equation $f(x) = 0$ has only one solution,

    c show that the solution of $f(x) = 0$ lies in the interval $(1.8, 1.9)$.

    The iterative formula $x_{n+1} = \sqrt{4 - \ln x_n}$ is used to solve the equation $f(x) = 0$.

    d Using $x_0 = 1.8$, calculate the values of x_1, x_2, x_3 and x_4. Hence find an approximation to the solution of $f(x) = 0$, giving your answer to 3 decimal places.

    Another attempt is made to find the solution of $f(x) = 0$ using the iterative formula $x_{n+1} = e^{4 - x_n^2}$ with $x_0 = 1.8$.

    e Describe the outcome of such an attempt.

Practice exam paper

Answer **all** questions.

Time allowed: 1 hour 30 minutes

A calculator **may** be used in this paper.

1 Express $\dfrac{x}{x+2} + \dfrac{3x-4}{x^2-x-6}$ as a single fraction in its simplest form. *(5 marks)*

2 The curve C has equation $y = x^2 \ln x$, $x > 0$. The point A, with x-coordinate e, lies on C. Find an equation of the tangent to C at A. *(8 marks)*

3 a Using the identities for $\sin(A + B)$ and $\cos(A + B)$, prove that

$$\tan 2\theta \equiv \frac{2\tan\theta}{1 - \tan^2\theta}.$$

Given that $\tan 2\theta = -\frac{1}{2}$, *(4 marks)*

b use the identity proved in **a** to find the two possible exact values of $\tan\theta$. *(4 marks)*

4 The diagram shows part of the curve with equation $y = \mathrm{f}(x)$, $x \in \mathbb{R}$. The curve passes through the origin and has a maximum point at $(2, 3)$.

On separate diagrams, sketch the curve with equation

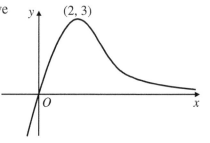

a $y = \mathrm{f}(x + 2)$, *(2 marks)*

b $y = \mathrm{f}(\frac{1}{2}x) + 2$, *(3 marks)*

c $y = \mathrm{f}(|x|)$. *(3 marks)*

In each case show the coordinates of any points at which the curve has a maximum point.

5 Solve, for $0 \leqslant \theta < 180$, the equation

$$4\operatorname{cosec}^2 2\theta° = 3 + 5\cot 2\theta°.$$ *(10 marks)*

6 $\qquad \mathrm{f}(x) = \sin(2x - 3) - \frac{1}{2}x + 1$,

where x is in radians.

a Show that there is a root α of $\mathrm{f}(x)$ in the interval $[1.32, 1.34]$. *(2 marks)*

b Show that the equation $\mathrm{f}(x) = 0$ can be written in the form

$$x = l\arcsin(\tfrac{1}{2}x - 1) + m,$$

stating the value of the constant l and the value of the constant m. *(3 marks)*

The root α is to be estimated using the iterative formula

$$x_{n+1} = l\arcsin(\tfrac{1}{2}x_n - 1) + m, \ x_0 = 1.33,$$

with the values of l and m found in **b**.

c Calculate x_1, x_2, x_3 and x_4, giving your answers to five decimal places. *(3 marks)*

d By considering the change of sign of $\mathrm{f}(x)$ over a suitable interval, show that, to four decimal places, $\alpha = 1.3289$. *(3 marks)*

7 The function f is defined by

$$f: x \mapsto \frac{3x - 2}{x + 2}, x > -2.$$

    a Find $f^{-1}(x)$.   *(3 marks)*

    b Write down the range of f^{-1}.   *(1 marks)*

The function g is defined by

$$g: x \mapsto \ln(2x + 4), x > -2.$$

    c Calculate gf(1), giving your answer to three significant figures.   *(3 marks)*

    d Solve $f^{-1}g(x) = 6$, giving your answer to three significant figures.   *(5 marks)*

8 The population P of slugs in a field t hours after a natural predator has been introduced into the field to control the slugs is given by

$$P = \frac{7000}{3e^{kt} + 4},$$

where k is a constant.

    a Write down the value of P when $t = 0$.   *(1 mark)*

Given that the initial population of slugs has been halved when $t = 60$,

    b calculate the value of k.   *(6 marks)*

    c Show that $\dfrac{dP}{dt} = -\dfrac{21\,000k\,e^{kt}}{(3e^{kt} + 4)^2}$.   *(4 marks)*

    d Using the value of k found in **b**, calculate the value of $\dfrac{dP}{dt}$ when $t = 60$.   *(2 marks)*

Answers

SKILLS CHECK 1A (page 3)

1 **a** $\dfrac{x}{3x-2}$ **b** $\dfrac{x+2}{x+5}$ **c** $\dfrac{2}{x-3}$ **d** $\dfrac{3x+1}{2x-3}$

2 **a** $x(x-7)$ **b** $\dfrac{x+4}{x}$ **c** $\dfrac{x-1}{2}$ **d** $\dfrac{t}{3}$

3 **a** $\dfrac{8x+7}{(x-1)(x+2)}$ **b** $\dfrac{14-3y}{4-y}$

 c $\dfrac{x-8}{(x-2)(x-5)}$ **d** $\dfrac{2x-3}{(x-3)(x-1)}$

4 $2x^2 - x + 6 - \dfrac{2x-7}{x^2-2}$

5 **a** $x^2 + x + 1$ **b** $x = -2$ or 1

SKILLS CHECK 1B (page 7)

1 **a** One-to-one function **b** Not a function
 c One-to-one function **d** Many-to-one function
2 **a** $f(x) = \{2.5, 3, 3.5, 4\}$ **b** $f(x) = \{\frac{1}{4}, \frac{1}{3}, \frac{1}{2}, 1\}$
 c $g(x) \in \mathbb{R},\ g(x) \geq 0$ **d** $f(x) \in \mathbb{R},\ f(x) \geq -2$
3 **a** **i** See CD. **b** **i** See CD.
 ii $f(x) \in \mathbb{R}$ **ii** $g(x) \in \mathbb{R}, -1 \leq g(x) \leq 1$
 c **i** See CD. **d** **i** See CD.
 ii $f(x) \in \mathbb{R}, f(x) > 0$ **ii** $h(x) \in \mathbb{R}, h(x) \geq -9$
4 $f(x) \in \mathbb{R}, 4\sqrt{3} \leq f(x) \leq 8$

SKILLS CHECK 1C (page 13)

1 **a** -13 **b** 4 **c** 12 **d** 7 **e** $3\frac{1}{4}$ **f** 6
2 **a** 2^{3x+2} **b** x **c** $\dfrac{3}{x}+2$ **d** $\dfrac{1}{3x+2}$
3 **a** -4.5 **b** -4 **c** $\pm 2\sqrt{2}$ **d** -3 or -6
4 **a** No inverse, many-to-one **b** Inverse, one-to-one
5 **a** **i** $f^{-1}: x \mapsto \dfrac{x-5}{2}$, domain $x \in \mathbb{R}$, range $f^{-1}(x) \in \mathbb{R}$
 ii See CD.
 b **i** $f^{-1}: x \mapsto 3 - 4x$, domain $x \in \mathbb{R}$, range $f^{-1}(x) \in \mathbb{R}$
 ii See CD.
 c **i** $f^{-1}: x \mapsto \sqrt{x}$, domain $x \in \mathbb{R}, x \geq 0$, range $f^{-1}(x) \in \mathbb{R}, f^{-1}(x) \geq 0$
 ii See CD.
 d **i** $f^{-1}: x \mapsto x^2 + 3, x \in \mathbb{R}, 0 \leq x \leq 3$, range $f^{-1}(x) \in \mathbb{R}, 3 \leq f^{-1}(x) \leq 12$
 ii See CD.
6 **a** $\frac{17}{19}$ **b** $\frac{1}{2}$ or -1 **c** $\frac{1}{3}$ or 1
7 **a** $f^{-1}: x \mapsto \dfrac{2}{3-x}, x \neq 3$ **b** 1 or 2
8 **a** $\frac{1}{4}$ or 1 **b** $f^{-1}: x \mapsto \dfrac{5x-1}{4x}, x \neq 0$ **c** -2

SKILLS CHECK 1D (page 21)

1 **a** See CD for graph; $(0, 3), (\frac{3}{2}, 0)$
 b See CD for graph; $(\frac{1}{2}, 0), (0, 1)$
 c See CD for graph; $(0, 4)$
2 **a** See CD for graph; $(-4, 0), (-2, -4), (0, 0)$
 See CD for graph; $(-4, 0), (-2, 4), (0, 0)$
 b See CD.
 c See CD for graph; $(0, 1), (\frac{1}{2}\pi, 0), (\pi, -1), (\frac{3}{2}\pi, 0), (2\pi, 1)$
 See CD for graph; $(0, 1), (\frac{1}{2}\pi, 0), (\pi, 1), (\frac{3}{2}\pi, 0), (2\pi, 1)$
3 **a** See CD for graph; $(a, 0), (0, a)$
 b See CD for graph; $\left(-\dfrac{a}{2}, 0\right), (0, a)$
 c See CD for graph; $(0, 2a)$

4 **a** See CD.
 b **i** $\frac{2}{3}$ or 4 **ii** 6 or $\frac{4}{5}$ **iii** $2a$ or $\dfrac{2a}{3}$
5 **a** Stretch in the y-direction by scale factor $\frac{1}{2}$, translation by -3 units in the y-direction.
 b Reflection in the y-axis, stretch in the y-direction by scale factor 0.4.
 c Translation by 2 units in the x-direction, stretch in the y-direction by scale factor 5.
6 **a** See CD for graph; $(-2, -2), (0, 1), (2, -2)$
 b See CD for graph; $(-2, 0), (0, -3), (2, 0)$
 c See CD for graph; $(0, 1), (1, 0)$
 d See CD for graph; $(3, 5), (5, 2)$
7 **a** $y = 3\tan x - 2$ **b** $y = -\sin\frac{1}{3}x$
8 **a** See CD.
 b See CD for graph; $(\frac{1}{4}\pi, 0), (\frac{3}{4}\pi, 1), (\frac{5}{4}\pi, 0), (\frac{7}{4}\pi, -1)$
 c $-1 \leq g(x) \leq 1$

Exam practice 1 (page 22)

1 $\dfrac{3x}{2x+1}$

2 **a** $\dfrac{21-x}{(x+3)(x-8)}$ **b** $9, -5$

3 **a** $\frac{1}{5}$ **b** $g^{-1}: x \mapsto \dfrac{1}{x}, x \in \mathbb{R}, x \neq 0$ **c** $-1, 2$
4 **a** $gf: x \mapsto 4 - 9x^2, x \in \mathbb{R}$ **b** See CD.
 c $gf(x) \in \mathbb{R}, gf(x) \leq 4$ **d** 0.4
5 **b** $f(x) \in \mathbb{R}, 0 < f(x) < \frac{4}{3}$
 c $f^{-1}(x) = \dfrac{4-x}{2x}, x \in \mathbb{R}, x > \frac{4}{3}$
 d $f^{-1}(x) \in \mathbb{R}, f^{-1}(x) > 1$
6 **a** $f^{-1}: x \mapsto \dfrac{x+1}{4}, x \in \mathbb{R}$
 b $gf: x \mapsto \dfrac{3}{8x-3}, x \in \mathbb{R}, x \neq \frac{3}{8}$
 c -0.076 (3 d.p.), 0.826 (3 d.p.)
7 **a** $g(x) \in \mathbb{R}, g(x) \geq 0$ **b** $0, 8$
 c See CD for graph; $2, 6$
8 **a** See CD. **b** $0, -4$
9 **a** $5a$ **b** See CD. **c** $0, -2a$
10 See CD.
11 See CD.
12 **a** $(0, 2)$ **b** $(0, 2), (0.5, 0)$ **c** $(0, -0.5), (1, 0)$
13 **a** See CD. **b** See CD. **c** $a = -2, b = -1$ **d** $x = \frac{1}{6}$

SKILLS CHECK 2A (page 31)

1 **a** $63°$ **b** $-42°$
 c $70°$ **d** $102°$
2 **a** $\frac{1}{4}\pi$ **b** $-\frac{1}{6}\pi$ **c** $\frac{1}{2}\pi$ **d** $\frac{1}{4}\pi$
3 **a** $23.6°$ (1 d.p.), $156.4°$ (1 d.p.) **b** $150°, 330°$
 c $24.1°$ (1 d.p.), $155.9°$ (1 d.p.), $204.1°$ (1 d.p.), $335.9°$ (1 d.p.)
 d $60°, 120°, 240°, 300°$
4 **a** 2.38^c (3 s.f.) **b** $\frac{3}{2}\pi$
 c 3.39^c (3 s.f.), 6.03^c (3 s.f.) **d** $\frac{1}{4}\pi, \frac{3}{4}\pi, \frac{5}{4}\pi, \frac{7}{4}\pi$
5 See CD.
6 $60°, 180°, 300°$
7 $-330°, -210°, 30°, 150°$
8 $\frac{1}{6}\pi, \frac{1}{2}\pi, \frac{5}{6}\pi, \frac{3}{2}\pi$
9 $-\frac{3}{4}\pi, \frac{1}{4}\pi$
10 **a** Translate by $-90°$ in the x-direction, reflect in the x-axis.
 b See CD. **c** They are the same curve.
11 **a** Stretch by scale factor 2 in the y-direction, translate by 1 unit in the y-direction.
 b $(\frac{1}{2}\pi, 3)$ **c** $f(x) \geq 3$
12 **a** Stretch in the x-direction by scale factor $\frac{1}{2}$, translate by 1 unit in the y-direction.
 b $(180, 2)$

SKILLS CHECK 2B (page 35)

1 a $\frac{3}{5}$ b $\frac{5}{13}$ c $\frac{56}{65}$ d $\frac{65}{56}$

2 b $45°, 215°$

3 See CD.

4 a $0°, 180°, 210°, 330°, 360°$ b $45°, 90°, 135°, 225°, 270°, 315°$

5 a $0, \frac{2}{3}\pi, \frac{4}{3}\pi, 2\pi$ b $0, \frac{4}{3}\pi$

6 b 0.13^c (2 d.p.), 1.44^c (2 d.p.) 3.27^c (2 d.p.), 4.59^c (2 d.p.)

7 a i $\cos 2A = 2\cos^2 A - 1$ ii $\cos 2A = 1 - 2\sin^2 A$

8 a $\sin(X - Y) = \sin X \cos Y - \cos X \sin Y$

9 $\frac{1}{2}$

10 See CD.

SKILLS CHECK 2C (page 39)

1 a $\sqrt{5} \cos(x - 26.6°)$ b $90°, 323.1°$ (1 d.p.)

2 a $f(\theta) = 5\cos(\theta + 36.9°)$ b i 5   ii $-36.9°$

3 a $\sqrt{2} \sin(x + \frac{1}{4}\pi)$ b $-\frac{1}{12}\pi$

4 a $5\sin(x - 53.1°)$

   b Translate by $53.1°$ in the x-direction and stretch by scale factor 5 in the y-direction.

5 $\alpha = 61.9°, k = 8$

Exam practice 2 (page 39)

1 $60°, 109.5°$ (1 d.p.), $250.5°$ (1 d.p.), $300°$

2 0.29^c (2 d.p.), 1.08^c (2 d.p.), 1.86^c (2 d.p.), 2.65^c (2 d.p.)

3 $\frac{1}{6}\pi, \frac{1}{2}\pi, \frac{5}{6}\pi$

4 $2.5°, 92.5°, 182.5°$

5 a $3\sin 2x - 2\cos 2x$

   b $16.8°$ (1 d.p.), $106.8°$ (1 d.p.), $196.8°$ (1 d.p.), $286.8°$ (1 d.p.)

6 a $\frac{1}{4}\pi, \frac{1}{2}\pi, \frac{3}{4}\pi, \frac{3}{2}\pi$ b $\frac{1}{3}\pi, \frac{5}{6}\pi, \frac{4}{3}\pi, \frac{11}{6}\pi$

7 b 1.355^c (3 d.p.), 4.497^c (3 d.p.)

8 See CD.

9 a $25\cos(\theta - 1.287\ldots^c)$ b 0.2^c (1 d.p.), 2.3^c (1 d.p.)

10 a $2\sin(x + 60°)$ b See CD.

   c See CD.   d $40°, 60°, 160°$

11 a $DX = 2\sin\theta \Rightarrow DE = 4\sin\theta$

   b Angle $BDG = \theta$ so $DG = EF = 2\cos\theta \Rightarrow P = 8\sin\theta + 4\cos\theta$

   c $4\sqrt{5}\sin(\theta + 0.463\ldots^c)$

   d 0.791^c (3 s.f.), 1.42^c (3 s.f.)

SKILLS CHECK 3A (page 46)

1 $x = 3\ln 2$

2 $x = \ln 2$

3 $x = \frac{5}{2}$

4 $x = \dfrac{e^{0.5} - 3}{4}$

5 See CD for graph; $(1, 0), (2, 0), (0, \ln 3)$

6 a i $f^{-1}(x) = \ln\dfrac{x}{3}$

     ii See CD for graph; $(0, 3), (3, 0)$

     iii Range of $f : f(x) \in \mathbb{R}, f(x) > 0$,

        domain of $f^{-1}(x) : x \in \mathbb{R}, x > 0$,

        range of $f^{-1}(x) : f^{-1}(x) \in \mathbb{R}$

   b i $f^{-1}(x) = \frac{1}{2}e^x$

     ii See CD for graph; $(\frac{1}{2}, 0), (0, \frac{1}{2})$

     iii Range of $f : f(x) \in \mathbb{R}$,

        domain of $f^{-1}(x) : x \in \mathbb{R}$,

        range of $f^{-1}(x) : f^{-1}(x) \in \mathbb{R}, f^{-1}(x) > 0$

7 a $f^{-1}(x) = \dfrac{\ln x - 1}{2}$

   b See CD for graph; asymptotes $x = 0, y = 0$

   c Range of $f : f(x) \in \mathbb{R}, f(x) > 0$,

     domain of $f^{-1}(x) : x \in \mathbb{R}, x > 0$,

     range of $f^{-1}(x) : f^{-1}(x) \in \mathbb{R}$.

8 a $f^{-1}(x) = e^{4x} - 1$

   b Domain of $f^{-1}(x) : x \in \mathbb{R}$,

     range of $f^{-1}(x) : f^{-1}(x) \in \mathbb{R}, f^{-1}(x) > -1$.

   c $e^2 - 1$

9 See CD.

10 a £1568.31   b 8 years   c See CD.

Exam practice 3 (page 47)

1 a $x = \ln 4$

   b $y = \dfrac{e}{e - 1}$

2 a $\ln 2 - 2$

   b $\frac{1}{2}e^3 + 2$

3 a See CD.   b $(-4, 0), (0, 8), (0, 2)$

4 a See CD for graph; $(0, e - 2), (\ln 2 - 1, 0)$

   b $f^{-1} x \mapsto \ln(x + 2) - 1$

   c domain of $f^{-1}(x) : x \in \mathbb{R}, x > -2$,

     range of $f^{-1}(x) : f^{-1}(x) \in \mathbb{R}$.

5 a See CD.   b $(2e^{-k}, 0)$

6 a $f^{-1}(x) = \dfrac{e^x + 6}{3}$

   b domain of $f^{-1}(x) : x \in \mathbb{R}$,

     range of $f^{-1}(x) : f^{-1}(x) \in \mathbb{R}, f^{-1}(x) > 2$.

   c 8.70

   d See CD.

   e $(0, \ln 6), (\frac{5}{3}, 0), (\frac{7}{3}, 0)$

7 a $f^{-1}(x) = \frac{1}{2}(4 - e^x)$

   b See CD for graph; $(0, \frac{3}{2}), (\ln 4, 0)$

   c 0.369

   d 2.141

8 a See CD.   b 14

SKILLS CHECK 4A (page 53)

1 a $5e^x$ b $\dfrac{3}{x}$ c $-4\sin x - \cos x$

   d $6x^2 - e^x$ e $-\dfrac{2}{x}$

2 a $-6x\sin x^2$ b $-6\sin x\cos x$ c $3(2x - 6)e^{x^2 - 6x}$

   d $\dfrac{8}{2x - 5}$ e $6\tan 3x \sec^2 3x$

3 $y = \ln 5 + \ln x - \frac{1}{3}\ln(2x + 7)$

   $\dfrac{dy}{dx} = \dfrac{4x + 21}{3x(2x + 7)}$

4 $\left(\dfrac{\pi}{6}, \dfrac{3}{2}\right)$, maximum

5 $y + 6x + 11 = 0$

6 $(2, 4 - 8\ln 2)$

7 a $3e^x - \dfrac{2}{x}$

   b $y + 2\ln 5 = 3e x - 2x + 2$

   c $2 - \ln 25$

8 $(0, 0)$

SKILLS CHECK 4B (page 56)

1 a $x^2(x + 4)(5x + 12)$ b $\sin x(\sec^2 x + 1)$

   c $2e^{2x}x^3(2 + x)$ d $\dfrac{x}{2x + 6} + \ln\sqrt{x + 3}$

2 a $\dfrac{3x(x - 6)}{(x - 3)^2}$ b $\dfrac{1}{1 - \sin x}$

   c $\dfrac{e^{\frac{x}{2}}(x - 6)}{4x^4}$ d $\dfrac{1 - \ln(x + 1)}{(x + 1)^2}$

3 a $(x^2 + 3)^2(5x - 4)^4(55x^2 - 24x + 75)$ b $3\cos^2 x \cos 4x$

   c $\dfrac{4e^x}{(2e^x + 1)^2}$ d $\dfrac{2(x\cos 2x - \sin 2x)}{x^3}$

4 $-e^{-3}$

5 $(0, 0), (3, 27e^{-3})$

6 b $\dfrac{18}{(x-3)^3}$

7 $2y + 4x = 5$

8 $\frac{1}{2}$

SKILLS CHECK 4C (page 58)

1 $\dfrac{1}{y}$

2 a $\dfrac{1}{\cos y}$ **b** $-\dfrac{1}{\sin y}$

3 $\dfrac{(y-1)^2}{y^2 - 2y - 1}$

4 $\dfrac{1}{4x^{\frac{3}{4}}}$

5 a $\dfrac{3}{y}$ **b** $6y + 2x = 7$

6 $\dfrac{3}{1 - \ln 3}$

7 a $y = x + 1, 2y = x + 4$ **b** $(2, 3)$

8 $12y = 13x - 25$

Exam practice 4 (page 59)

1 Tangent is $y - 4 + \ln 5 = 5(x - 1)$

2 $y = 2x + 4$

3 a $q = 2.5, p = -0.5$

4 a i $x = a^y$

    c $y - 1 = \dfrac{1}{10 \ln 10}(x - 10)$ **d** $10 - 10 \ln 10$

5 b $\dfrac{x^2 + 6x + 7}{(x + 3)^2}$ **c** $x = -7$

7 i $3x^2 e^{3x} + x^3 e^{3x}$ **ii** $\dfrac{2 \cos x + 2x \sin x}{\cos^2 x}$

    iii $2 \tan x \sec^2 x$ **iv** $-\dfrac{1}{2y \sin y^2}$

8 a $T = 80$ **b** $e^{-0.1t} > 0$

    c See CD.   **d** 4.1

    e $-6e^{-0.1t}$ **f** 38

9 a $p = \frac{1}{2} \ln 5, q = -\frac{4}{3}$ **b** $9y + 6x + 8 = 0$

10 a $2y \ln y + y$ **b** $\dfrac{1}{3e}$

SKILLS CHECK 5A (page 62)

1 a $f(5) = -0.290..., f(6) = 0.817...$

    b $f(-1) = -1.41..., f(0) = 1$

    c $f(4.1) = -0.277..., f(4.2) = 0.439...$

    d $f(-4) = -1.17..., f(-3) = 0.092...$

    e $f(1.2) = 0.105..., f(1.3) = -0.427...$

2 a See CD.   **b** Two

    c $f(2) = 0.693..., f(3) = -3.90...$

3 a See CD.   **c** $n = 1$

SKILLS CHECK 5B (page 65)

1 a ii $1.5275..., 1.5196..., 1.5218..., 1.521$

    b ii $2.25129..., 2.25156..., 2.25162..., 2.2516$

    c ii $-0.338295..., -0.339630..., -0.339680..., -0.33968$

    d ii $1.2164..., 1.2173..., 1.2174..., 1.217$

    e ii $0.2116..., 0.2143..., 0.2150..., 0.215$

2 a i $1.3234..., 1.3240..., 1.324$

    ii $f(1.3235) = -0.00953..., f(1.3245) = 0.00421...$

b i $-2.1966..., -2.1960..., -2.196$

    ii $f(-2.1955) = 0.00370..., f(-2.1965) = -0.00776...$

c i $0.1704..., 0.1686..., 0.169$

    ii $f(0.1685) = -0.00288..., f(0.1695) = 0.00403...$

d i $11.02, 11.0164..., 11.016$

    ii $f(11.0155) = -0.119..., f(11.0165) = 0.00249...$

e i $7.5078..., 7.5070..., 7.507$

    ii $f(7.5065) = -0.000245..., f(7.5075) = 0.00160...$

3 a $f(1) = 2.54..., f(2) = -1.41...$

    b $1.6944..., 1.6960..., 1.696$

4 a See CD.

    c $2.28462..., 2.28128..., 2.28056...$

    d 2.28

5 a $f(5) = -3, f(6) = 4$

    b $7, 10.25, 24.265...$

    c Diverges

Exam practice 5 (page 66)

1 a $f(0.5) = 0.125, f(0.6) = -0.450...$

    b $0.5182, 0.5180, 0.5180$

    c $f(0.51795) = 0.000328..., f(0.51805) = -0.000336...$

2 a $a = 0.1, b = -2$

    b $-1.104165..., -1.104538..., -1.104572...$

    c -1.105

3 a $f(2) = 0.210..., f(3) = -0.735...$

    b $2.189647..., 2.189586..., 2.189576...$

    c 2.1896

    d $f(2.18955) = 0.0000240..., f(2.18965) = -0.0000759...$

4 a $6x - e^{-x}$

    b $f'(0.1) = -0.304..., f'(0.2) = 0.381...$

    c $0.1508..., 0.1433..., 0.1444..., 0.144$

5 a $f(1.1) = -0.0648... = -0.065$ (2 s.f.),

      $f(1.15) = 0.09894... = 0.099$ (2 s.f.)

    c $p = \frac{4}{5}, q = \frac{2}{5}$

      $1.112965..., 1.117610..., 1.119245..., 1.120$

6 b $1.32327..., 1.32653..., 1.327$

    c $p = -1.25$

    d $-2.63017..., -2.63933..., -2.642$

7 a $f(1) = -2, f(2) = 5\frac{1}{2}$

    b $1.38672..., 1.39609..., 1.39527..., 1.39534...,$ root $= 1.395$

    c $f(1.3945) \approx -0.005, f(1.3955) \approx 0.001$

8 a See CD

    c $f(1.8) = -0.172..., f(1.9) = 0.251...$

    d $1.8472..., 1.8401..., 1.8412..., 1.8410...,$ approx solution $= 1.841$

    e $2.138..., 0.564..., 39.709...$ attempt fails; sequence does not converge

Practice exam paper (page 68)

1 $\dfrac{x - 2}{x - 3}$

2 $y = 3ex - 2e^2$

3 b $2 \pm \sqrt{5}$

4 a See CD for graph; $(0, 3)$

    b See CD for graph; $(4, 5)$

    c See CD for graph; $(-2, 3), (2, 3)$

5 $\theta = 22\frac{1}{2}, 112\frac{1}{2}; \theta \approx 38, 128$

6 b $l = \frac{1}{2}, m = \frac{3}{2},$

    c $x_1 = 1.32920, x_2 = 1.32898, x_3 = 1.32893, x_4 = 1.32891$

7 a $f^{-1}(x) = \dfrac{2 + 2x}{3 - x}$

    b $f^{-1}(x) > -2$

    c 1.54

    d 1.69

8 a 1000 **b** $k \approx 0.02$ **c** $\dfrac{dP}{dt} \approx -7.2$